Practice Papers for SQA Exams

National 5

Physics

001/30072014

10 9 8 7 6 5 4 3

ISBN 9780007504749

Published by
Leckie & Leckie Ltd
An imprint of HarperCollins*Publishers*
Westerhill Road, Bishopbriggs, Glasgow, G64 2QT
T: 0844 576 8126   F: 0844 576 8131
leckieandleckie@harpercollins.co.uk   www.leckieandleckie.co.uk

Publisher: Peter Dennis
Project Manager: Craig Balfour

Special thanks to
Jill Laidlaw (copy edit); Donna Cole (proofread);
Ink Tank (cover design); QBS (layout)

Printed by CPI Group (UK) Ltd, Croydon, CR0 4YY

A CIP Catalogue record for this book is available from the British Library.

Acknowledgements

Images: P93 © Bloomberg via Getty Images.
All other images © Shutterstock.com.

Illustrations © HarperCollins Publishers

Whilst every effort has been made to trace the copyright holders, in cases where this has been unsuccessful, or if any have inadvertently been overlooked, the Publishers would gladly receive any information enabling them to rectify any error or omission at the first opportunity.

Introduction

The three papers included in this book are designed to provide practice in the National 5 Physics Course assessment question paper (the examination), which is worth 80% of the final grade for this course.

Together, the three papers give overall and comprehensive coverage of the assessment of **Knowledge and its Application** as well as the **Problem Solving Skills** needed to pass National 5 Physics. The **Key Area Index** grid on page 5 shows the pattern of coverage of the knowledge in the Key Areas and the problem solving skills required across the three papers.

We recommend that candidates download a copy of the Course Assessment Specification from the SQA website at www.sqa.org.uk. Print pages 9–12, which summarise the knowledge and skills that will be tested.

Design of the papers

Each paper has been carefully assembled to be very similar to a typical National 5 question paper. Each paper has 110 marks and is divided into two sections.

- **Section 1** – Objective Test, which contains 20 multiple choice items worth 1 mark each and totalling 20 marks altogether.
- **Section 2** – Paper 2, which contains restricted and extended response questions totalling 90 marks altogether. During the marking process this will be scaled to 60 marks.

In each paper, the marks are distributed evenly across all three component units of the course and the majority of the marks are awarded for applying knowledge and understanding. The other marks are for applying scientific inquiry, scientific analytical thinking and problem solving skills.

A data sheet containing relevant data is provided with each paper along with a relationship sheet containing a list of all formulae. Candidates should familiarise themselves with the data and formulae that will be made available to them during the exam. Data and relationship sheets can be found on pages 7 and 8 respectively in this book and should be consulted when required.

Most questions in each paper are set at the standard of Grade C but there are also more difficult questions set at the standard for Grade A. We have attempted to construct each paper to represent the typical range of demand in a National 5 Physics paper.

Using the papers

Each paper can be attempted as a whole or as groups of questions on a particular topic or skill area – use the **Key Area Index** grid to find related groups of questions. Open-ended questions and problem solving questions are also highlighted in the index grid. In the grid, questions may appear more than once if they cover more than one skill area. Use the Date completed column to keep a record of your progress.

We recommend attempting the questions before studying their Expected Answers.

Use a **pen**, **sharp pencil**, **clear plastic ruler** and a **calculator** for the best results. A couple of different coloured highlighters could also be handy.

Expected answers

The expected answers on pages 95–128 give National Standard answers but, occasionally, there may be other acceptable answers. For example, when giving a numerical answer there will be a range of significant figures that will be acceptable. As a rough guide, try and give answers to a similar number of significant figures as the data given in the question.

There are Top Tips provided alongside each answer. These include hints on the physics itself as well as some memory ideas, a focus on traditionally difficult areas, advice on the wording of answers and notes of commonly made errors.

Grading

Exams A, B and C are designed to be equally demanding and to reflect the National Standard of a typical SQA paper. Each paper has 110 marks – if you score 55 marks, that's a C pass. You will need about 66 marks for a B pass, and about 77 marks for an A. These figures are a rough guide only.

Timing

If you are attempting a full paper, limit yourself to **2 hours** to complete it. We recommend around 30 minutes for the Objective Test and the remainder of the time for Section 2.

For extended response questions give yourself about a minute per mark, for example, a 10-mark question should take no longer than around 10 minutes.

Good luck!

Topic index

Skill tested	Key area	Practice paper questions S1 – Section 1 S2 – Section 2			Date completed
		Exam A	**Exam B**	**Exam C**	
Unit 1: Electricity and Energy Demonstrating and applying knowledge	1. Conservation of energy	**S1:** **S2:** 8(c)	**S1:** – **S2:** 1, 2(c)	**S1:** **S2:** 2(a), 2(b)	
	2. Electrical charge carriers and electric fields	**S1:** 2	**S1:** 1	**S1:** 2 **S2:** 1(b)	
	3. Potential difference (voltage)	**S1:** 3			
	4. Ohm's law	**S1:** – **S2:** 1(a), 1(b)	**S1:** 2 **S2:** 3(a)	**S1:** – **S2:** 1(a)(ii)	
	5. Practical electrical and electronic circuits	**S1:** 4, 5	**S1:** 3, 4 **S2:** 3(b), 3(c)	**S1:** 4, 6 **S2:** 1(a)(i)	
	6. Electrical power		**S1:** 5 **S2:** 2(a)	**S1:** 5	
	7. Specific heat capacity	**S1:** 6	**S1:** – **S2:** 2(b)	**S1:** 3	
	8. Gas laws and the kinetic model	**S1:** 7, 8 **S2:** 2(a), 2(b)	**S1:** 6, 7, 8	**S1:** 7, 8 **S2:** 3	
Unit 2: Waves and Radiation Demonstrating and applying knowledge	1. Wave parameters and behaviours	**S1:** 9, 10 **S2:** 5	**S1:** 9, 10 **S2:** 6(b), 6(c)	**S1:** 9, 10 **S2:** 6	
	2. Electromagnetic spectrum	**S1:** 11 **S2:** 4(b)	**S1:** 11 **S2:** 4(b), 6(a), 7(b)	**S1:** 11 **S2:** 7(a)	
	3. Light	**S1:** 12, 13	**S1:** 12 **S2:** 7(a)	**S1:** 12, 13	
	4. Nuclear radiation	**S1:** 14, 15 **S2:** 6, 7	**S1:** 13, 14, 15 **S2:** 8	**S1:** 14, 15 **S2:** 8	

Continued . . .

Unit 3: Dynamics and Space Demonstrating and applying knowledge	1. Velocity and displacement – vectors and scalars	**S1:** 16, 17 **S2:** 9(d)	**S1:** 16, 17	**S1:** 16 **S2:** 9	
	2. Velocity-time graphs	**S1:** 18 **S2:** 8(a), 8(b)	**S1:** 18	**S1:** – **S2:** 10(a), 10(b)	
	3. Acceleration			**S1:** 17	
	4. Newton's laws	**S1:** 19, 20 **S2:** 9(a), 9(b)(i), 9(b)(iii), 9(c)	**S1:** 19, 20 **S2:** 10	**S1:** 18, 20 **S2:** 10(c)	
	5. Projectile motion	**S1:** – **S2:** 9(b)(ii)	**S1:** – **S2:** 9	**S1:** 19 **S2:** 7(b)	
	6. Space exploration			**S1:** – **S2:** 2(a)	
	7. Cosmology	**S1:** – **S2:** 10(a), 10(b)	**S1:** – **S2:** 11	**S1:** – **S2:** 11	
N5 Physics Course Problem solving skills		**S1:** 5, 7, 17, 18 **S2:** 1(a)(ii), 2(b), 4(a), 6(a)(i), 7(c), 9(b)(iii), 10(b), 10(c)	**S1:** 4, 5, 11 **S2:** 1(b), 2(c), 3(c)(ii), 4(a), 4(b), 6(d), 8(a), 10(b)	**S1:** 6, 7, 12, 17 **S2:** 1(b), 2(a), 3(c), 4, 7(b), 8(b), 9(c), 11(b)	
N5 Physics Course Open-ended questions		**S2:** 3, 11	**S2:** 5, 12	**S2:** 5, 12	

Data Sheet

Speed of light in materials

Material	Speed in m s^{-1}
Air	$3 \cdot 0 \times 10^8$
Carbon dioxide	$3 \cdot 0 \times 10^8$
Diamond	$1 \cdot 2 \times 10^8$
Glass	$2 \cdot 0 \times 10^8$
Glycerol	$2 \cdot 1 \times 10^8$
Water	$2 \cdot 3 \times 10^8$

Gravitational field strengths

	Gravitational field strength on the surface in N kg^{-1}
Earth	9·8
Jupiter	23.0
Mars	3·7
Mercury	3·7
Moon	1·6
Neptune	11.0
Saturn	9·0
Sun	270.0
Uranus	8·7
Venus	8·9

Specific latent heat of fusion of materials

Material	Specific latent heat of fusion in J kg^{-1}
Alcohol	$0 \cdot 99 \times 10^5$
Aluminium	$3 \cdot 95 \times 10^5$
Carbon Dioxide	$1 \cdot 80 \times 10^5$
Copper	$2 \cdot 05 \times 10^5$
Iron	$2 \cdot 67 \times 10^5$
Lead	$0 \cdot 25 \times 10^5$
Water	$3 \cdot 34 \times 10^5$

Specific latent heat of vaporisation of materials

Material	Specific latent heat of vaporisation in J kg^{-1}
Alcohol	$11 \cdot 2 \times 10^5$
Carbon Dioxide	$3 \cdot 77 \times 10^5$
Glycerol	$8 \cdot 30 \times 10^5$
Turpentine	$2 \cdot 90 \times 10^5$
Water	$22 \cdot 6 \times 10^5$

Speed of sound in materials

Material	Speed in m s^{-1}
Aluminium	5200
Air	340
Bone	4100
Carbon dioxide	270
Glycerol	1900
Muscle	1600
Steel	5200
Tissue	1500
Water	1500

Specific heat capacity of materials

Material	Specific heat capacity in J kg^{-1} °C^{-1}
Alcohol	2350
Aluminium	902
Copper	386
Glass	500
Ice	2100
Iron	480
Lead	128
Oil	2130
Water	4180

Melting and boiling points of materials

Material	Melting point in °C	Boiling point in °C
Alcohol	−98	65
Aluminium	660	2470
Copper	1077	2567
Glycerol	18	290
Lead	328	1737
Iron	1537	2737

Radiation weighting factors

Type of radiation	Radiation weighting factor
alpha	20
beta	1
fast neutrons	10
gamma	1
slow neutrons	3

Relationship Sheet

$E_p = mgh$

$E_k = \frac{1}{2}mv^2$

$Q = It$

$V = IR$

$R_T = R_1 + R_2 + ...$

$\frac{1}{R_T} = \frac{1}{R_1} + \frac{1}{R_2} + ...$

$V_2 = \left(\frac{R_2}{R_1 + R_2}\right)V_s$

$\frac{V_1}{V_2} = \frac{R_1}{R_2}$

$P = \frac{E}{t}$

$P = IV$

$P = I^2R$

$P = \frac{V^2}{R}$

$E_h = cm\Delta T$

$p = \frac{F}{A}$

$\frac{pV}{T} = \text{constant}$

$p_1V_1 = p_2V_2$

$\frac{p_1}{T_1} = \frac{p_2}{T_2}$

$\frac{V_1}{T_1} = \frac{V_2}{T_2}$

$d = vt$

$v = f\lambda$

$T = \frac{1}{f}$

$A = \frac{N}{t}$

$D = \frac{E}{m}$

$H = Dw_R$

$\dot{H} = \frac{H}{t}$

$s = vt$

$d = \bar{v}t$

$s = \bar{v}t$

$a = \frac{v - u}{t}$

$W = mg$

$F = ma$

$E_w = Fd$

$E_h = ml$

Practice Exam A

N5 Physics

Practice Papers for SQA Exams

Physics Section 1

Fill in these boxes:

Name of centre

Town

Forename(s)

Surname

Try to answer all of the questions in the time allowed.

Total marks — 110

Section 1 — 20 marks

Section 2 — 90 marks

Read all questions carefully before attempting.

You have 2 hours to complete this paper.

Write your answers in the spaces provided, including all of your working.

SECTION 1

Objective Test

1. A car of mass 1000 kg is travelling at a speed of 30 m s^{-1}. The brakes are applied and the car decelerates to 10 m s^{-1}.

The change in the car's kinetic energy is

A 10 kJ

B 200 kJ

C 400 kJ

D 445 kJ

E 800 kJ.

2. An electric heater has a rating of 1·25 kW, 50 Ω.

The charge passing through the element of the heater in 100 s is

A 2·5 C

B 50 C

C 500 C

D 2500 C

E 25 000 C.

3. The voltage of an electricity supply is a measure of the

A power developed in the circuit

B current in the circuit

C resistance of the circuit

D energy given to the charges in the circuit

E speed of the charges in the circuit.

4. Three resistors are connected as shown.

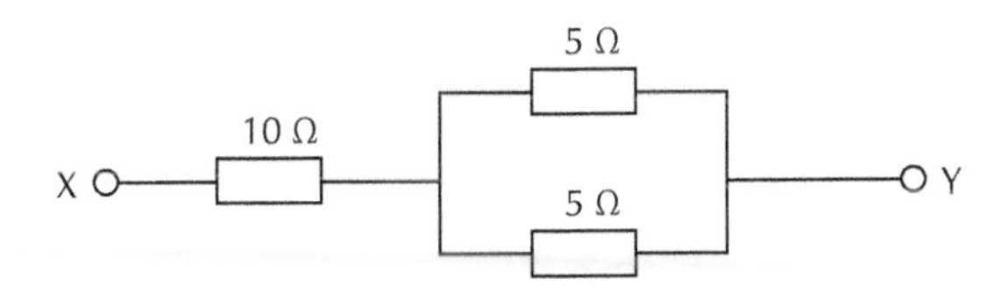

The resistance between X and Y is

A 2·00 Ω

B 10·4 Ω

C 12·5 Ω

D 15·0 Ω

E 20·0 Ω.

5. Two resistors are connected in series with a 54 volt d.c. supply.

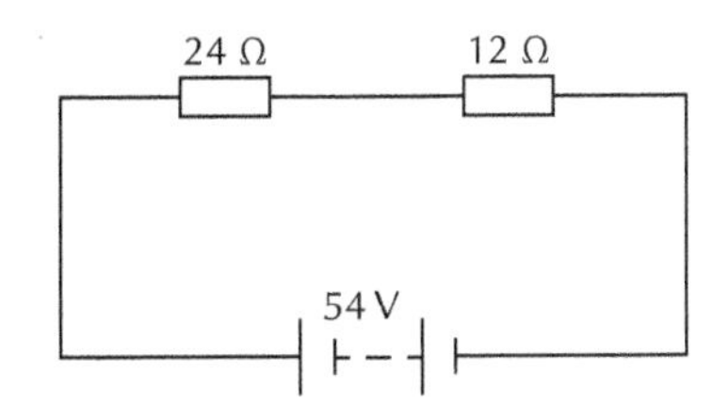

The current in the 24 Ω resistor is 1·5 A.

Which row of the table shows the current in the 12 Ω resistor and the voltage across the 12 Ω resistor?

	Current **(A)**	***Voltage*** **(V)**
A	0·75	9
B	1·5	18
C	1·5	54
D	3·0	36
E	3·0	54

6. A solar furnace holds 150 kg of water.

The water is heated from 25 °C to 85 °C.

The specific heat capacity of water is 4180 J kg^{-1} $°C^{-1}$.

The heat energy gained by the water is

A 9000 J

B 627 000 J

C 15 675 000 J

D 37 620 000 J

E 53 295 000 J.

7. The pressure–volume graph below describes the behaviour of a constant mass of gas when it is heated.

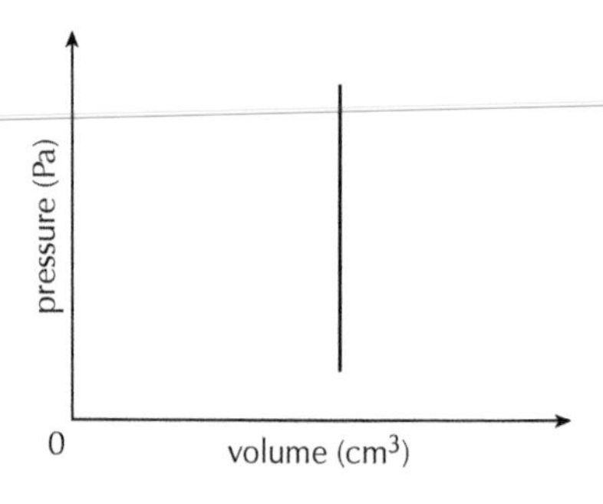

Which of the following shows the corresponding pressure–temperature graph?

A

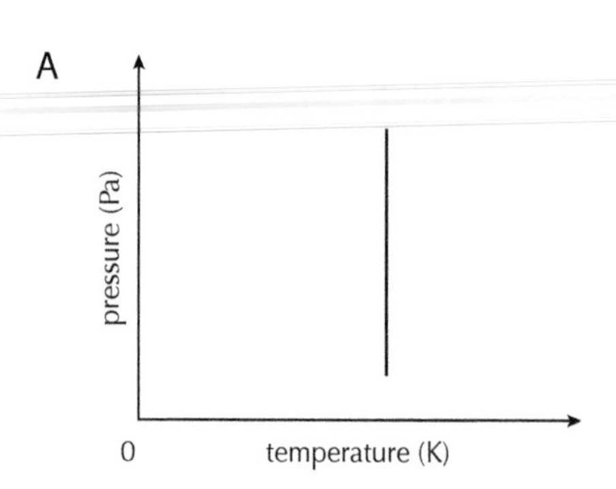

B

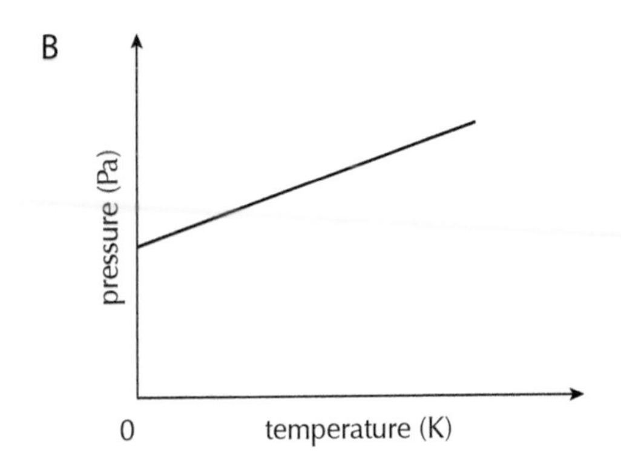
pressure (Pa)
0
temperature (K)

C

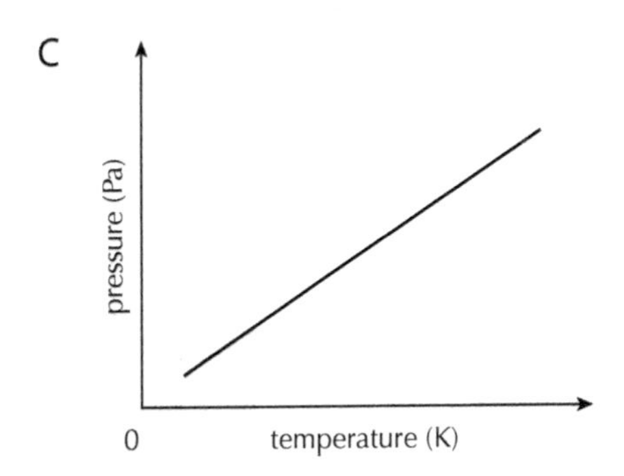
pressure (Pa)
0
temperature (K)

D

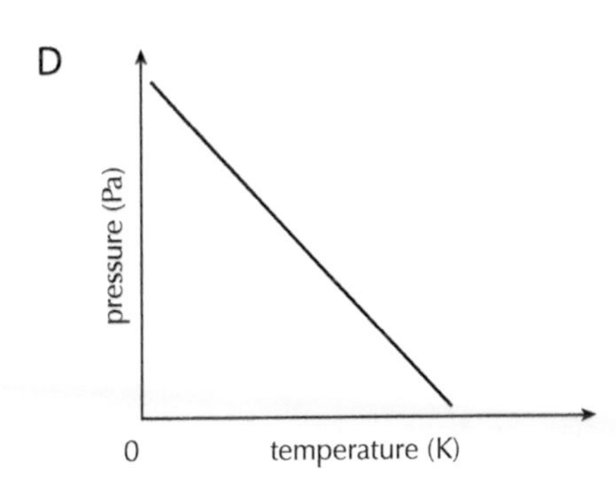
pressure (Pa)
0
temperature (K)

E

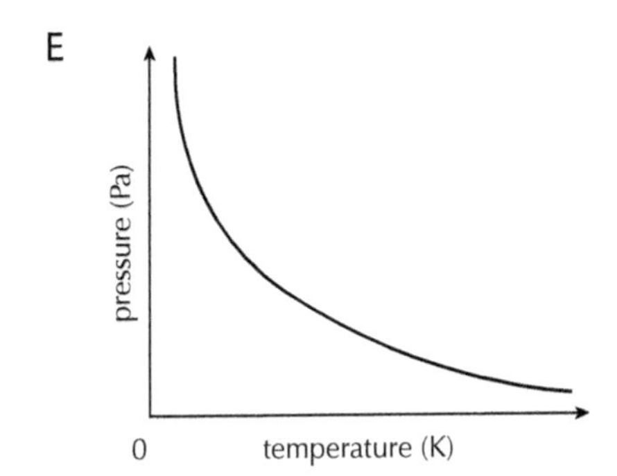
pressure (Pa)
0
temperature (K)

8. Water from a kettle at 90 °C cools to a temperature of 50 °C.

The temperature change on the Kelvin scale is

A 40 K

B 50 K

C 313 K

D 323 K

E 363 K.

9. A radio signal is sent from a transmitter in Scotland to a receiver in Canada.

The distance between transmitter and receiver is $5{\cdot}4 \times 10^3$ km. The signal has a frequency of 100 MHz.

The time taken for the signal to reach the receiver is

A $1{\cdot}8 \times 10^{-8}$ s

B $1{\cdot}8 \times 10^{-5}$ s

C $1{\cdot}6 \times 10^{-2}$ s

D $1{\cdot}8 \times 10^{-2}$ s

E $6{\cdot}0 \times 10^{-2}$ s.

10. In a swimming pool, 15 waves pass a point in 3 seconds. The speed of the waves is $0{\cdot}6$ m s^{-1}. The wavelength of the waves is

A 0·04 m

B 0·12 m

C 0·20 m

D 1·80 m

E 3·00 m.

11. The diagram shows part of the electromagnetic spectrum.

P	Microwaves	Q	Visible light

Identify radiation P and radiation Q.

	P	***Q***
A	radio	ultraviolet
B	radio	infrared
C	infrared	ultraviolet
D	ultraviolet	infrared
E	ultraviolet	radio

12. For a ray of light travelling at an angle from **glass into air**, which of the following statements is/are **true**?

I The speed of light increases.

II The wavelength of light increases.

III The direction of light changes.

A I only

B III only

C I and II only

D I and III only

E I, II and III

13. The diagram shows a ray of light striking a rectangular glass block.

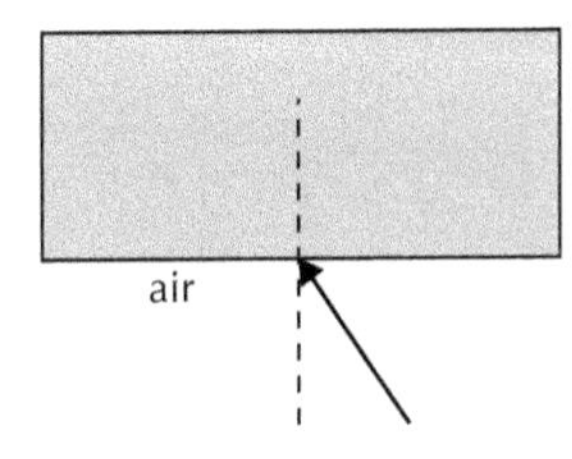

Which diagram shows the path of the ray through the block?

A

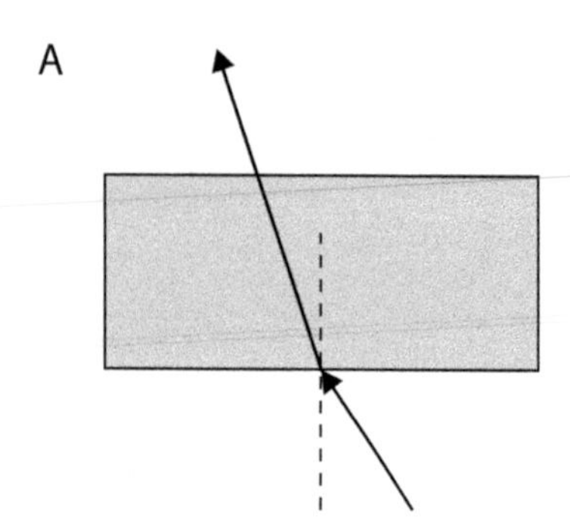

B

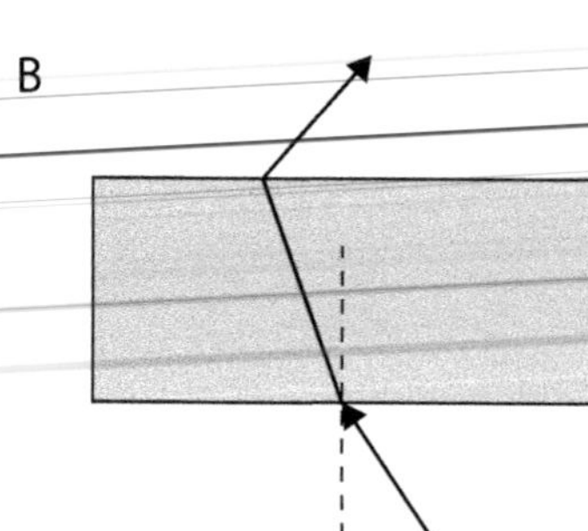

C

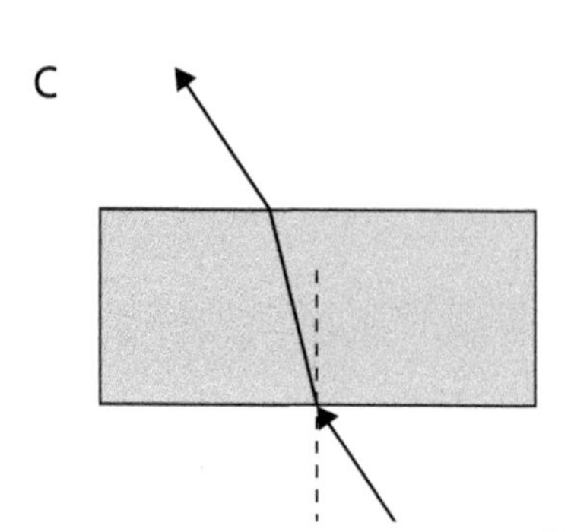

D

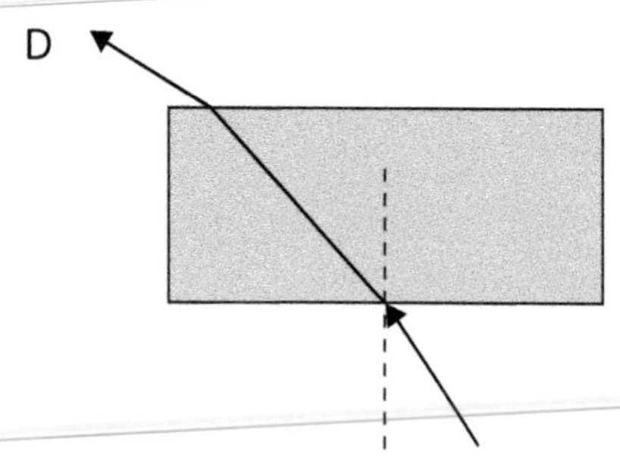

E

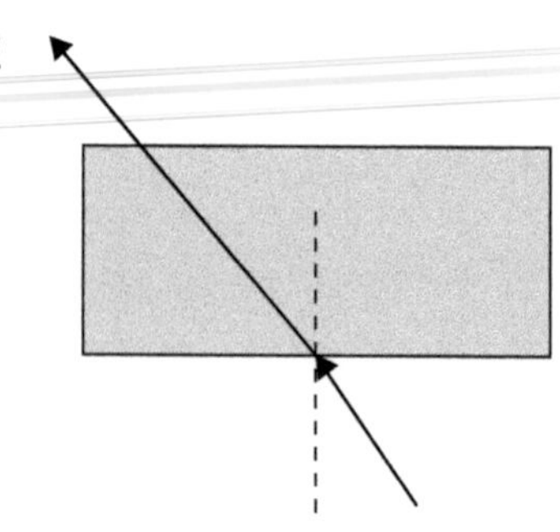

14. A student makes the following three statements.

I Alpha particles have a greater ionisation density than beta particles.

II Alpha particles have a greater ionisation density than gamma rays.

III Gamma rays have a greater ionisation density than alpha particles and beta particles.

Which of the statements is/are **true**?

A I only

B II only

C III only

D I and II only

E I and III only

15. The activity of a radioactive material is 180 Bq. The half-life of the substance is 8 hours.

The time for the activity to fall to 22·5 Bq is

A 4 hours

B 8 hours

C 16 hours

D 20 hours

E 24 hours.

16. Which of the following contains two vector quantities?

A Distance and speed

B Displacement and speed

C Displacement and velocity

D Weight and mass

E Force and mass

17. A student sets up the apparatus as shown.

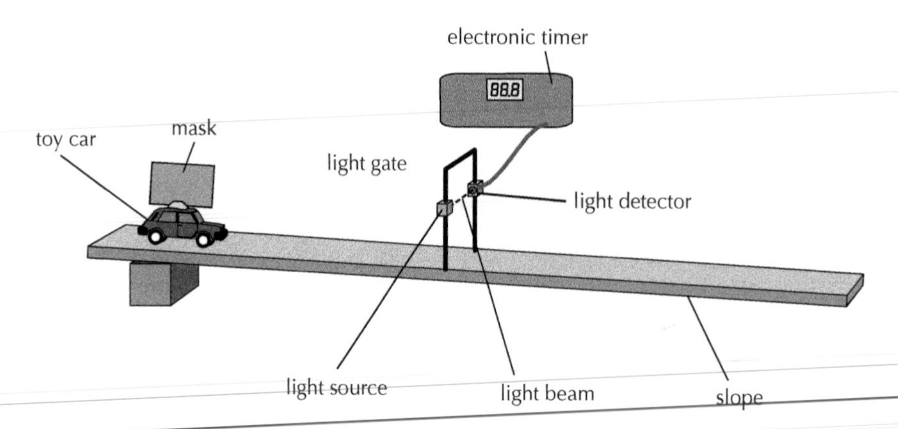

The toy car is released from rest at the top of the slope.

The following measurements are recorded.

Time for mask to pass through light gate = 0·04 s.

Time taken for toy car to travel from top of slope to light gate = 0·25 s.

Length of mask = 50 mm.

The instantaneous speed of the toy car as it passes through the light gate is

A $0{\cdot}20\ \text{m s}^{-1}$

B $0{\cdot}80\ \text{m s}^{-1}$

C $1{\cdot}25\ \text{m s}^{-1}$

D $5{\cdot}00\ \text{m s}^{-1}$

E $20{\cdot}0\ \text{m s}^{-1}$.

18. Three objects, X, Y and Z, travel in a straight line. The table below shows the velocities of the three objects for a time interval of 3 seconds.

***Time* (s)**	0	1	2	3
***Velocity of X* (m s^{-1})**	0	1	2	3
***Velocity of Y* (m s^{-1})**	0	1	3	5
***Velocity of Z* (m s^{-1})**	4	6	8	10

Which of the following statements is/are correct?

I X moves with constant acceleration.

II Y moves with constant acceleration.

III Z moves with constant velocity.

A I only

B II only

C III only

D I and III only

E I, II and III

19. An astronaut has a mass of 70 kg. Which row of the table gives the mass and weight of the astronaut on the Moon?

	Mass **(kg)**	***Weight*** **(N)**
A	112	112
B	112	70
C	70	70
D	70	112
E	70	686

20. A person sits on a chair that rests on the Earth. The person exerts a downward force on the chair.

Which of the following completes the 'Newton pair' of forces?

A The force of the person on the Earth.

B The force of the person on the chair.

C The force of the chair on the person.

D The force of the chair on the Earth.

E The force of the Earth on the person.

N5 Physics

Practice Papers for SQA Exams

Physics Section 2

Fill in these boxes:

Name of centre

Town

Forename(s)

Surname

Try to answer all of the questions in the time allowed.

Total marks — 110

Section 1 — 20 marks

Section 2 — 90 marks

Read all questions carefully before attempting.

You have 2 hours to complete this paper.

Write your answers in the spaces provided, including all of your working.

MARKS Do not write in this margin

SECTION 2

1. A student sets up the following circuit.

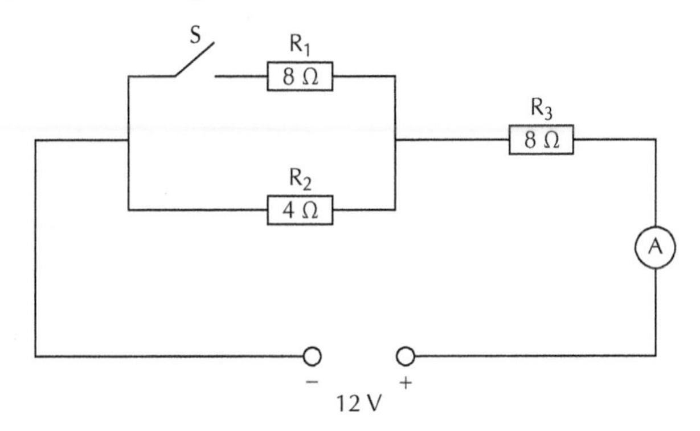

(a) (i) Calculate the current in the circuit when switch S is open. **4**

Space for working and answer

RT = R1 + R2
RT = 12 Ω

I = V/R
I = 12/12
I = 1 A

(ii) Calculate the potential difference across R_2 when switch S is closed. **5**

Space for working and answer

MARKS
Do not write in this margin

(b) The supply voltage is increased to 22 V. R_1 and R_2 are replaced by a 6 V lamp and switch S is removed. The lamp is operating at its stated voltage.

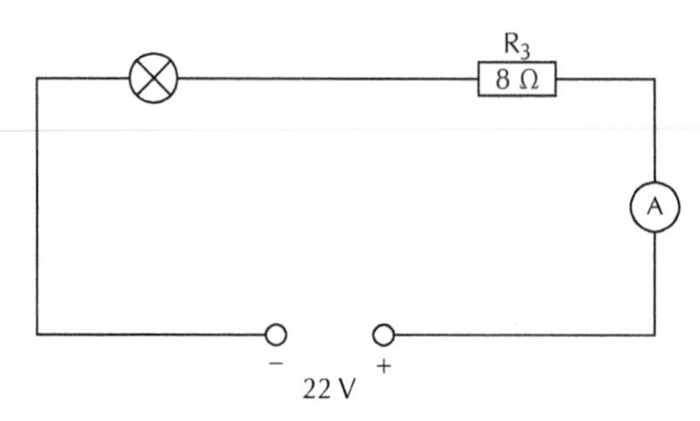

State whether there is a greater power dissipated in R_3 or in the lamp.

Justify your answer by calculation.

Space for working and answer **5**

Total marks 14

MARKS
Do not write in this margin

2. A fire extinguisher of compressed carbon dioxide gas is inside a van.

(a) The carbon dioxide inside the fire extinguisher is at a pressure of $2{\cdot}36 \times 10^6$ Pa and a temperature of 6·0 °C. The cylinder is now moved to a school laboratory where the temperature is 21·0 °C.

(i) Calculate the pressure of the carbon dioxide in the fire extinguisher when its temperature is 21·0 °C. **4**

Space for working and answer

(ii) Use the kinetic model to explain the change in pressure as the temperature increases. **2**

(b) The manufacturer warns against storing the fire extinguisher at temperatures above 48·0 °C.

Explain why a temperature limit must be set. **2**

Total marks 8

MARKS Do not write in this margin

3. Using your knowledge of physics, estimate the pressure you exert on the ground when standing on both feet. **3**

MARKS
Do not write in this margin

4. Manufacturers of solar cells need to know how efficient they are.

Efficiency is defined as the rate at which energy from the Sun is converted to useful output energy.

The efficiency of a solar cell can be determined using:

$$\textit{Percentage efficiency} = \frac{V_{oc} I_{sc}}{P_{in}} \times 100\%$$

where V_{oc} is the open circuit voltage, I_{sc} is the short circuit current and P_{in} is the input power.

The table below gives information on three different solar cells.

	V_{oc} (V)	I_{sc} (A)	P_{in} (W)
Solar cell A	0·7	3·8	10
Solar cell B	0·6	3·5	10
Solar cell C	0·98	2·3	15

(a) Show, by calculation, which solar cell has the highest percentage efficiency. **6**

Space for working and answer

(b) Solar cells use the infrared radiation from sunlight to generate power.

The wavelength of the radiation is 545 nm. Calculate the frequency. **3**

Space for working and answer

Total marks 9

MARKS
Do not write in this margin

5. An oven is used to cook popcorn. The oven produces microwaves.

(a) What is transferred by waves? **1**

(b) (i) Are microwaves transverse or longitudinal waves? **1**

(ii) Describe the difference between transverse and longitudinal waves. **2**

(c) Microwaves can also be used to transmit mobile telephone signals.

Complete the diagram below to show the pattern of the microwaves to the right of the hill. **2**

Total marks 6

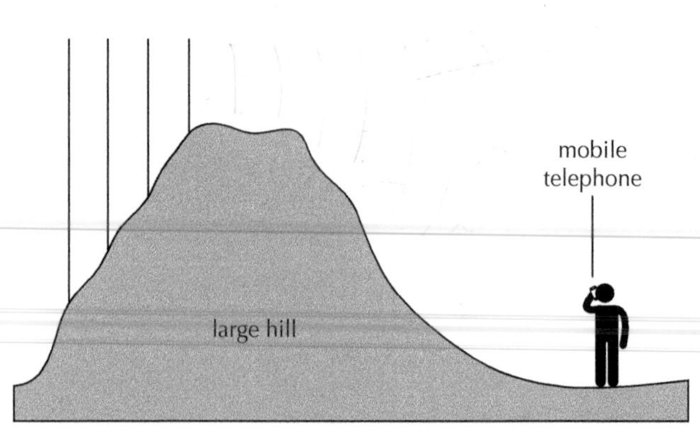

6. A school technician sets up the apparatus below to measure the half-life of a radioactive sample.

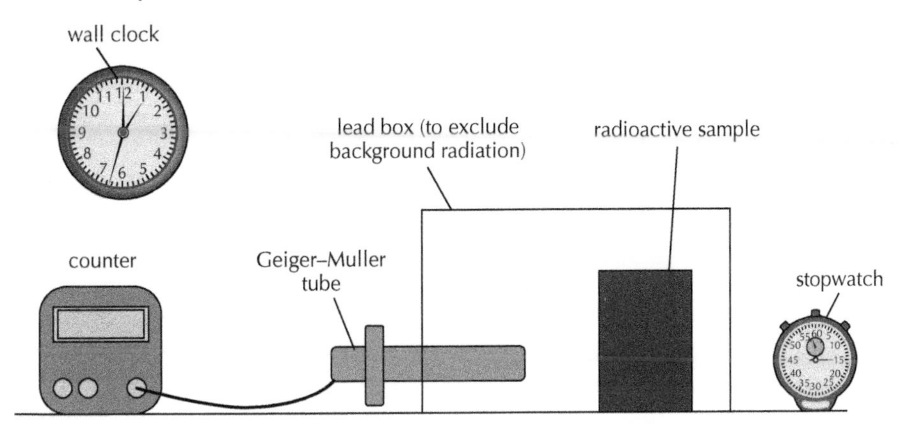

(a) (i) Describe how the technician could use the equipment to measure the half-life of the radioactive sample. Your description should include

- the apparatus required
- the measurements taken
- how the half-life is calculated. **3**

MARKS Do not write in this margin

(ii) A graph of count rate against time for the source is shown. The count rate has been corrected for background radiation.

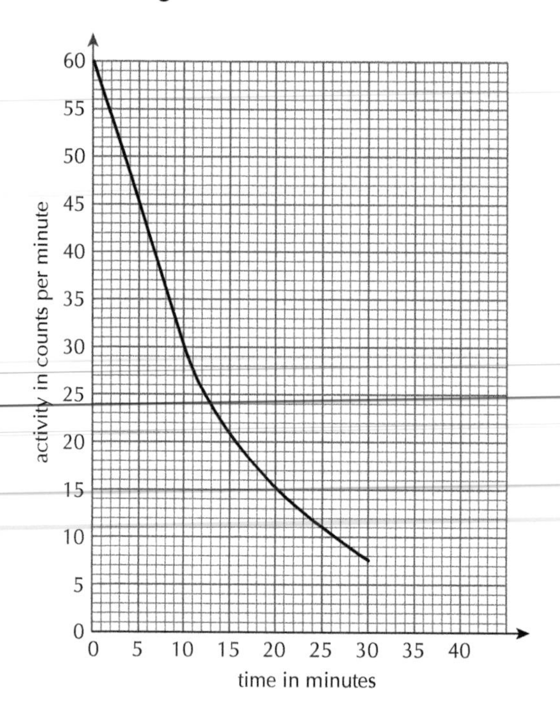

Use the graph to determine the half-life of the radioactive sample.

Space for working and answer **2**

(b) State **two** factors that can affect the background radiation level. **2**

(c) The radioactive sample emits alpha particles. What is an alpha particle? **1**

Total marks 8

MARKS
Do not write in this margin

7. Nuclear power stations can be used to produce electricity for the National Grid.

In the nuclear reactor of a power station, a neutron strikes the nucleus of a uranium atom. The uranium nucleus splits into two smaller parts and three neutrons are released.

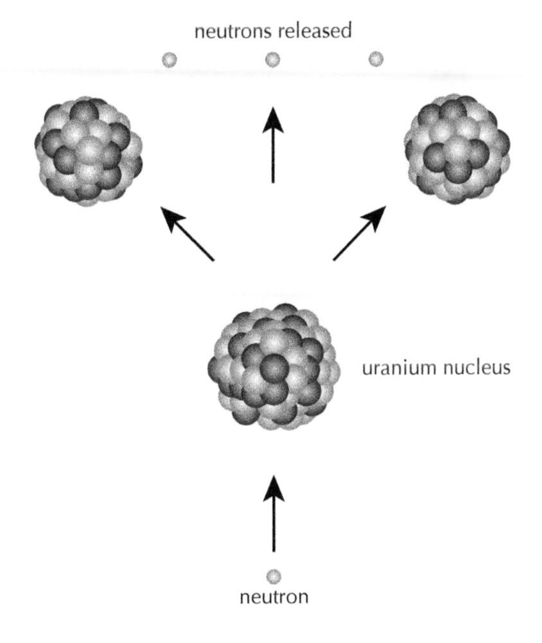

(a) State the name of this type of nuclear reaction. **1**

MARKS Do not write in this margin

(b) A worker at the nuclear power station is responsible for removing the control rods from the nuclear reactor.

The worker's hand receives an average absorbed dose of 0·06 μGy each time a control rod is handled. The radiation weighting factor of the radiation is 20.

Calculate the equivalent dose received by the worker.

Space for working and answer **3**

(c) In nuclear reactions, mass is converted to energy. The energy, in Joules, released in a nuclear reaction can be calculated using the equation:

$$E = mc^2$$

where m is the mass converted to energy in kilograms and c is the speed of light in air in metres per second.

The total mass before the nuclear reaction is $387{\cdot}497 \times 10^{-27}$ kg. The total mass after the nuclear reaction is $386{\cdot}822 \times 10^{-27}$ kg.

Calculate the energy released in the reaction.

Space for working and answer **4**

Total marks 8

MARKS Do not write in this margin

8. A weather balloon of mass 100 kg rises vertically from the ground.

The graph shows how the vertical velocity of the balloon changes during the first 120 seconds of its upward flight.

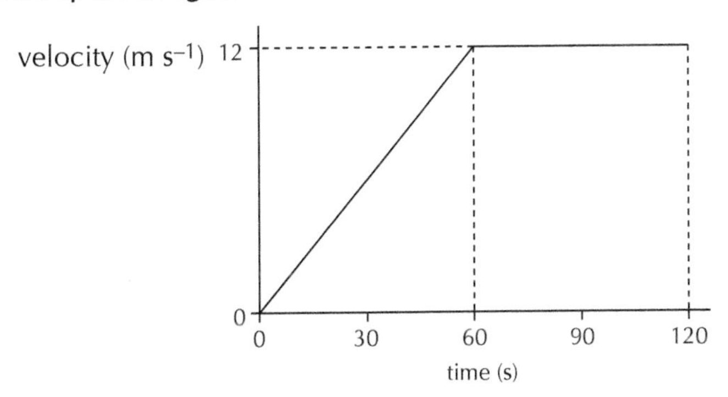

(a) Calculate the initial acceleration of the balloon. **3**

Space for working and answer

(b) Calculate the vertical displacement of the balloon after 120 s. **3**

Space for working and answer

(c) Calculate the kinetic energy of the balloon at 120 s. **3**

Space for working and answer

Total marks 9

MARKS Do not write in this margin

9. An aircraft is transporting skydivers to carry out a parachute jump. The aircraft and passengers have a total mass of 30 000 kg.

The forces exerted on the aircraft are shown on the diagram.

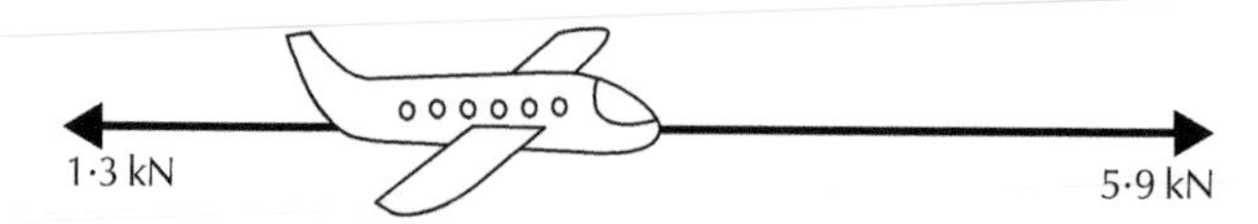

(a) Calculate the acceleration of the aircraft. **3**

Space for working and answer

(b) The aircraft reaches a constant horizontal speed. At a height of 5000 m a skydiver jumps from the aircraft. The skydiver falls for 24 seconds before opening her parachute.

(i) Sketch a diagram showing the forces acting on the skydiver as she falls through the air. You must name the forces and show their direction. **2**

(ii) Calculate the speed of the skydiver the instant before she opens her parachute. Assume the acceleration is constant. **3**

Space for working and answer

MARKS Do not write in this margin

(iii) Explain why the actual speed of the parachutist is much less than the value calculated in part (b) (ii). **2**

(c) A short time after opening her parachute, the skydiver reaches a terminal velocity. Explain what is meant by **terminal velocity**. **1**

(d) Shortly before landing, the skydiver is travelling at 4·8 m s^{-1} horizontally and 2·2 m s^{-1} vertically.

By scale diagram, or otherwise, determine the resultant velocity of the skydiver.

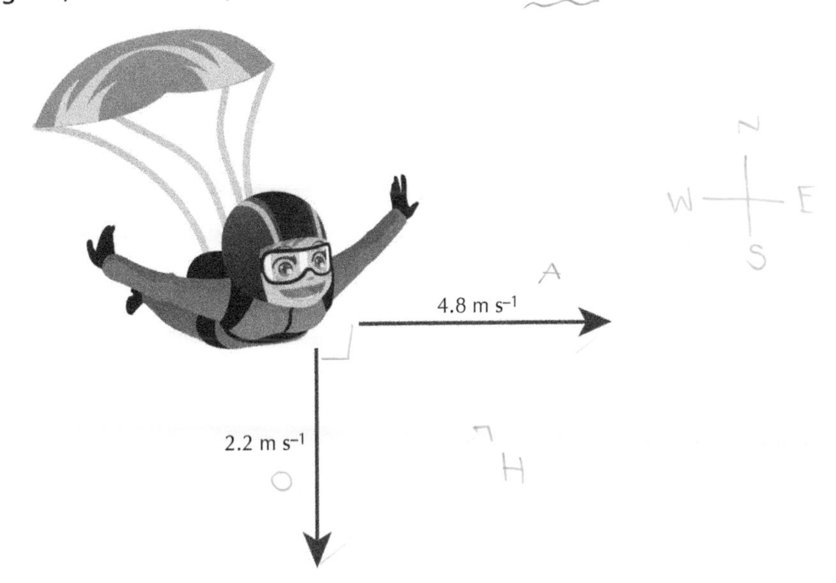

Space for working and answer **4**

$c^2 = a^2 + b^2$

$c = \sqrt{27.88}$

$c = 5.3 ms^{-1}$

$\tan = O/A$

$\tan^{-1} = 2.2/4.8$

$= 24.623$

↳ 5.3ms^{-1} at 24.6° horizontally vertical

Total marks 15

MARKS
Do not write in this margin

10. *Voyager 1* is a space probe launched by NASA in 1977 to explore the outer planets of the solar system.

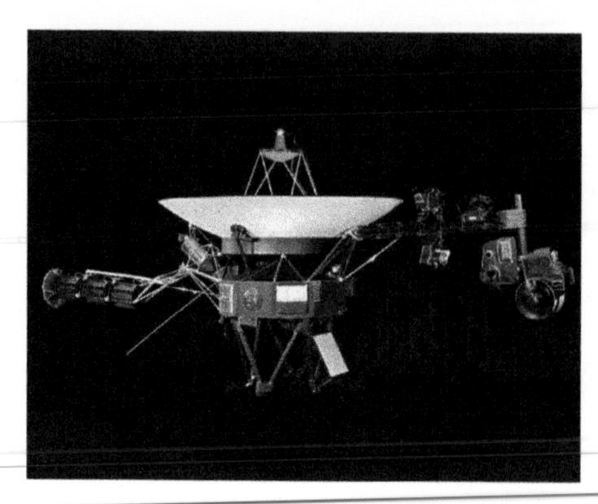

NASA communicates with the space probe via radio waves. In 2013, *Voyager 1* became the first man-made object to leave the solar system. The distance from Earth to *Voyager 1* is $1{\cdot}9 \times 10^{10}$ km.

(a) What is a planet? **1**

(b) A radio signal is transmitted from Earth to *Voyager 1*. The signal is processed and 5 minutes later is transmitted back to Earth. Calculate the total time taken for the signal to be sent and received back on Earth.

Space for working and answer **4**

(c) Distances in space are sometimes measured in astronomical units (AU).

1 AU = 149 597 871 km

Calculate the distance from Earth to Voyager 1 in astronomical units.

Space for working and answer **2**

Total marks 7

MARKS
Do not write in this margin

11. A science fiction film features an astronaut who becomes detached from a space shuttle in orbit around the Earth. The astronaut spins vertically away from the Earth into the vacuum of space.

Using your knowledge of physics, comment on the motion of the astronaut after becoming detached from the space shuttle. **3**

Space for working and answer

[END OF QUESTION PAPER]

Practice Exam B

N5 Physics

Practice Papers for SQA Exams

Physics Section 1

Fill in these boxes:

Name of centre	Town

Forename(s)	Surname

Try to answer all of the questions in the time allowed.

Total marks — 110

Section 1 — 20 marks

Section 2 — 90 marks

Read all questions carefully before attempting.

You have 2 hours to complete this paper.

Write your answers in the spaces provided, including all of your working.

SECTION 1

Objective Test

1. Electric current is the flow of

A Protons

B Neutrons

C Electrons

D Electrons and neutrons

E Protons and neutrons.

2. A range of potential differences are applied across a resistor and the corresponding currents are measured.

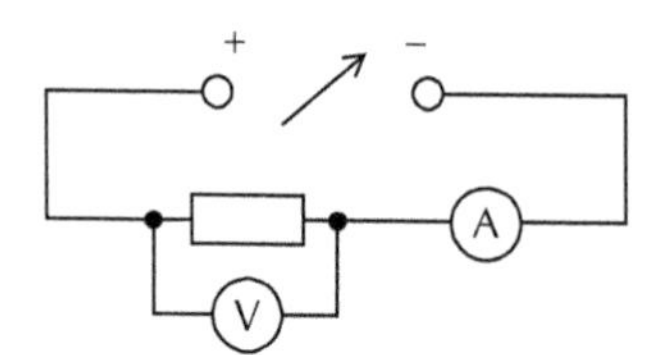

Assuming the temperature of the resistor remains constant, which of the following shows the current-voltage graph obtained?

A
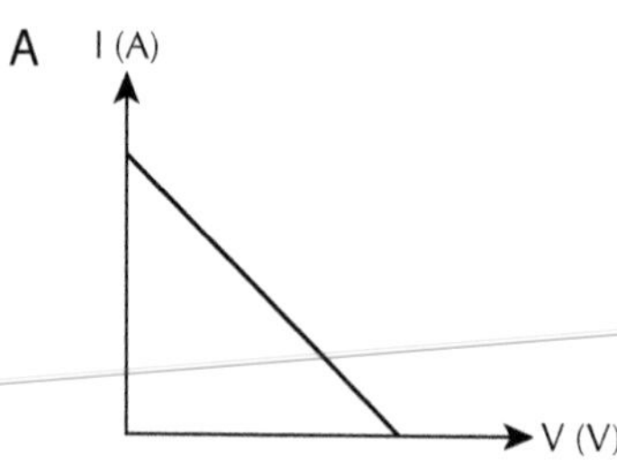

B
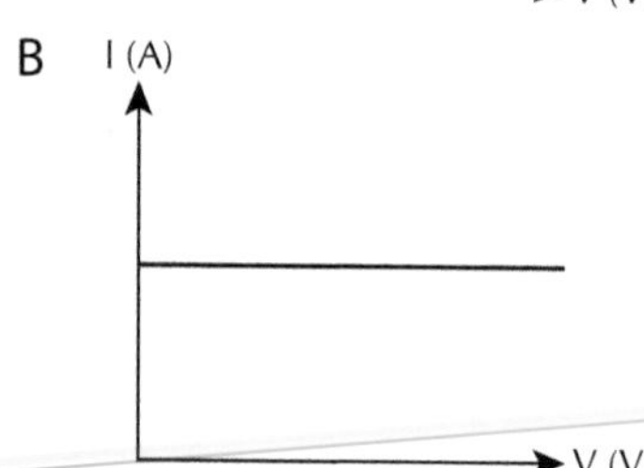

C
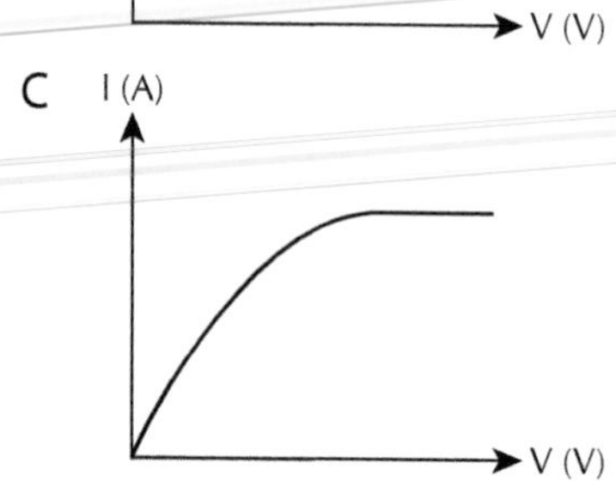

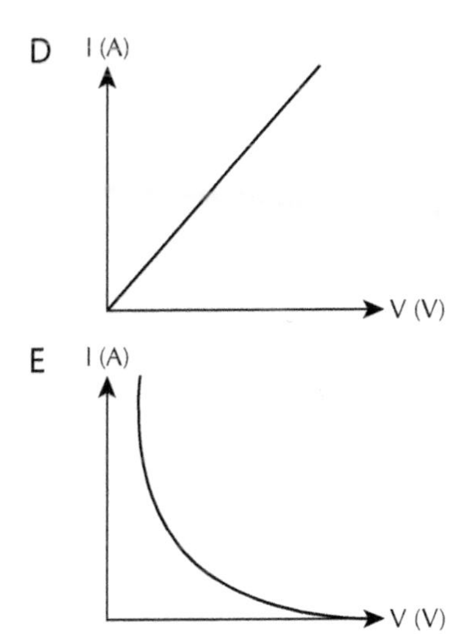

3. Consider the following circuit.

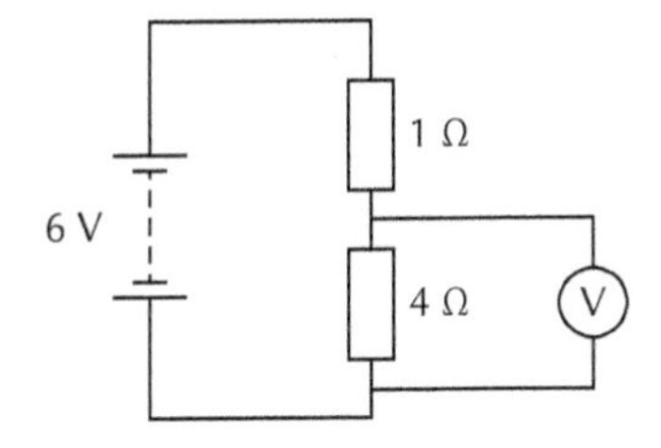

The reading on the voltmeter is

A 1·0 V

B 1·2 V

C 4·8 V

D 5·0 V

E 6·0 V.

4. Two identical resistors are connected with four ammeters as shown.

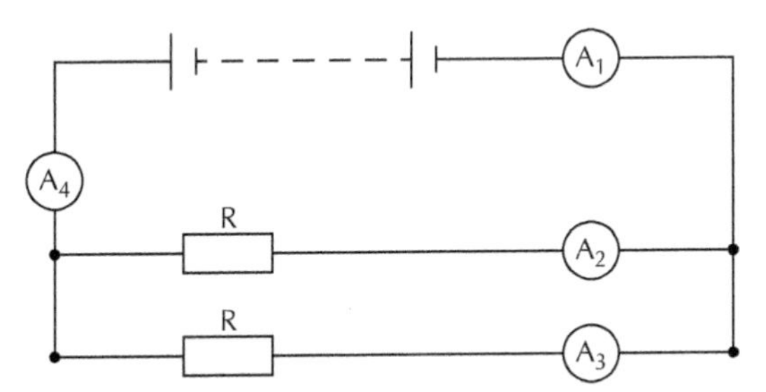

The reading on A_2 is 0·2 A.

Which row shows the readings on A_1, A_3 and A_4?

	Ammeter A_1	***Ammeter A_3***	***Ammeter A_4***
A	0·2 A	0·2 A	0·2 A
B	0·2 A	0·4 A	0·6 A
C	0·4 A	0·4 A	0·2 A
D	0·4 A	0·2 A	0·6 A
E	0·4 A	0·2 A	0·4 A

5. Which of the following equations can be used to find the power dissipated by a resistor?

I $P = IV$

II $P = I^2 R$

III $P = \frac{V^2}{R}$

A I only

B II only

C III only

D II and III only

E I, II and III

6. An aircraft is travelling at a constant height above the Earth. The air pressure at this height is $0{\cdot}2 \times 10^5$ Pa. The inside of the aircraft is maintained at a pressure of $1{\cdot}0 \times 10^5$ Pa.

The area of an external door is 2 m². What is the outward force produced on the door?

A $0{\cdot}4 \times 10^5$ N

B $0{\cdot}5 \times 10^5$ N

C $1{\cdot}6 \times 10^5$ N

D $2{\cdot}0 \times 10^5$ N

E $2{\cdot}4 \times 10^5$ N

7. The pressure of a fixed mass of gas is 200 kPa at a temperature of −17 °C. The volume of the gas remains constant.

At what temperature would the pressure of the gas be 300 kPa?

A 111 °C

B 162 °C

C 299 °C

D 384 °C

E 435 °C

8. The end of a syringe is sealed. The air inside is now trapped.

The plunger is pushed in slowly. The pressure of the air increases.

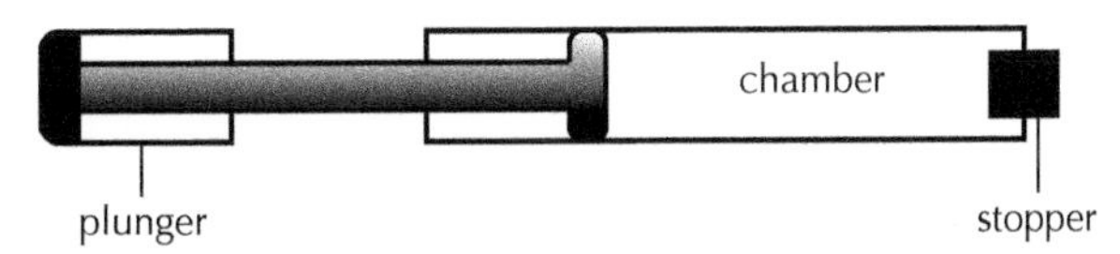

Which of the following explain(s) why the pressure increases, assuming the temperature remains constant?

I The air particles collide more often with the walls of the syringe.

II The air particles increase their average speed.

III The air particles strike the walls of the syringe with greater force.

A I only

B II only

C III only

D I and II only

E I and III only

9. Which of the following is a longitudinal wave?

A Water wave

B Light wave

C Sound wave

D Gamma ray

E Radio wave

10. A water wave is shown below.

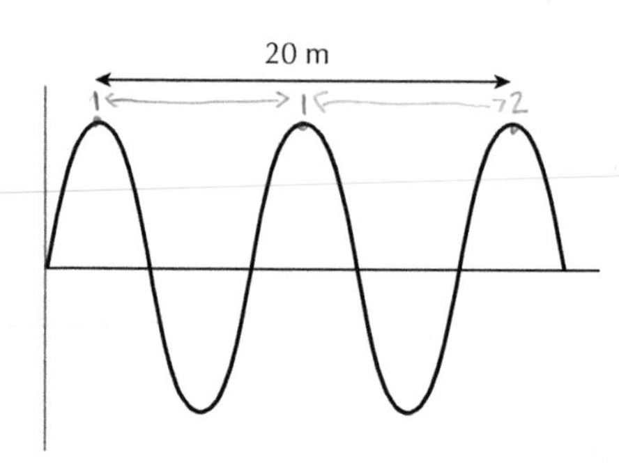

The speed of the wave is $2{\cdot}5\ \mathrm{m\ s^{-1}}$.

The frequency of the wave is

A 0·125 Hz

B 0·25 Hz

C 4·0 Hz

D 8·0 Hz

E 25 Hz.

11. Which of the following electromagnetic waves has a lower frequency than infrared and a shorter wavelength than radio?

A X-rays

B Gamma rays

C Visible light

D Ultraviolet

E Microwave

12. A ray of light is incident on a glass block as shown.

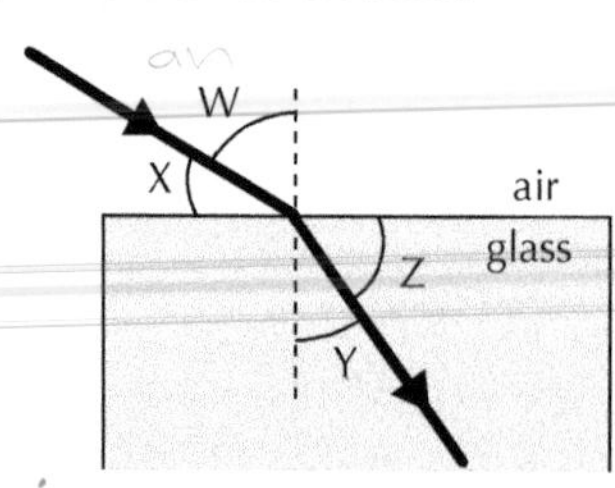

Which row correctly identifies the angle of incidence and the angle of refraction?

	Angle of incidence	*Angle of refraction*
A	W	Y
B	W	Z
C	X	Y
D	X	Z
E	Y	W

13. The diagram below shows a simple model of the atom.

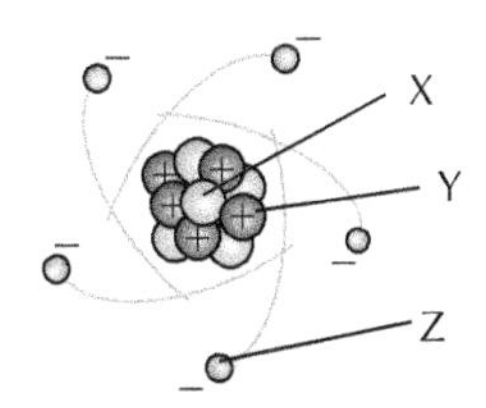

Which row in the table identifies particles X, Y and Z?

	X	*Y*	*Z*
A	neutron	proton	electron
B	proton	neutron	electron
C	neutron	electron	proton
D	proton	electron	neutron
E	electron	proton	neutron

14. A patient's thyroid gland has a mass of 0·04 kg.

The gland is exposed to radiation and absorbs 0·2 μJ of energy in 3 minutes.

The absorbed dose is

A 0·005 Gy

B 0·12 Gy

C 5 μGy

D 5 Gy

E 15 MGy.

15. The following statements are made about nuclear fission.

I The nucleus splits into smaller parts.

II Energy is released.

III Nuclei join together.

Which of the statements is/are correct?

A I only

B II only

C III only

D I and II only

E II and III only

16. Which row contains vector quantities only?

A	speed	distance	time
B	speed	displacement	acceleration
C	velocity	force	displacement
D	velocity	acceleration	displacement
E	velocity	distance	acceleration

17. A walker follows the route shown in the diagram.

Which row in the table shows the total distance travelled and the magnitude of the final displacement?

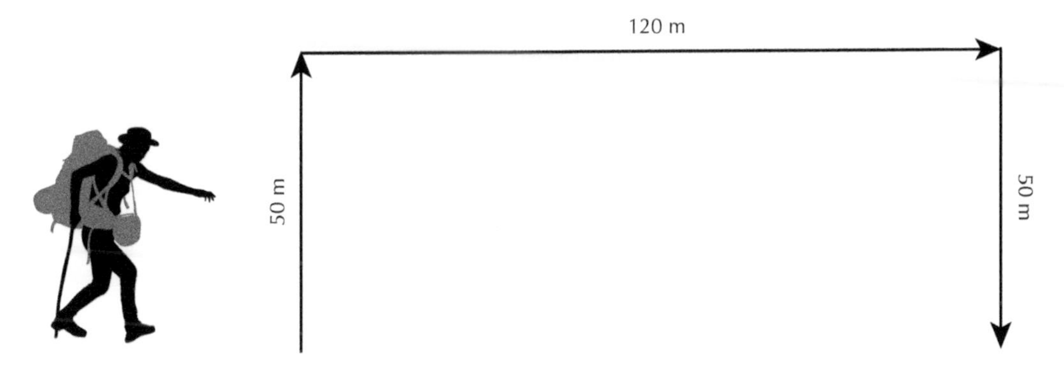

	Total distance **(m)**	***Final displacement*** **(m)**
A	120	50
B	120	220
C	170	120
D	220	120
E	220	220

18. A bus slows down as it approaches a set of traffic lights. The speed-time graph of the bus's motion is shown.

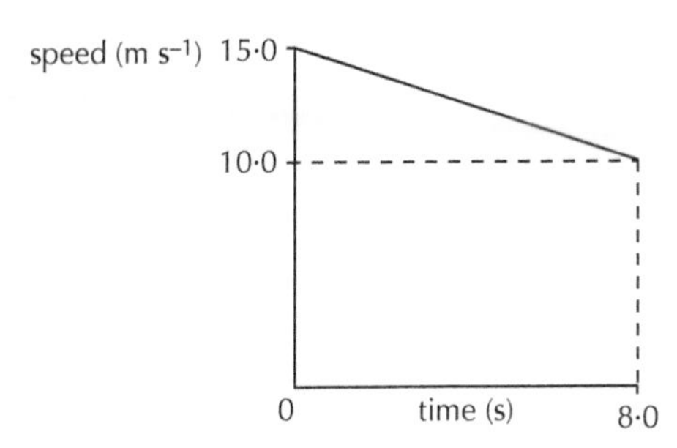

The distance travelled by the bus is

A 80 m

B 100 m

C 120 m

D 140 m

E 160 m.

19. A student writes the following statements about mass and weight.

I The mass of an object is different on Mars than it is on Earth.

II The mass of an object is the same on Mars as it is on Earth.

III The weight of an object is the same on Mars as it is on Earth.

Which of these statements is/are correct?

A I only

B II only

C III only

D I and III only

E II and III only

20. A toy car is travelling along a horizontal surface as shown.

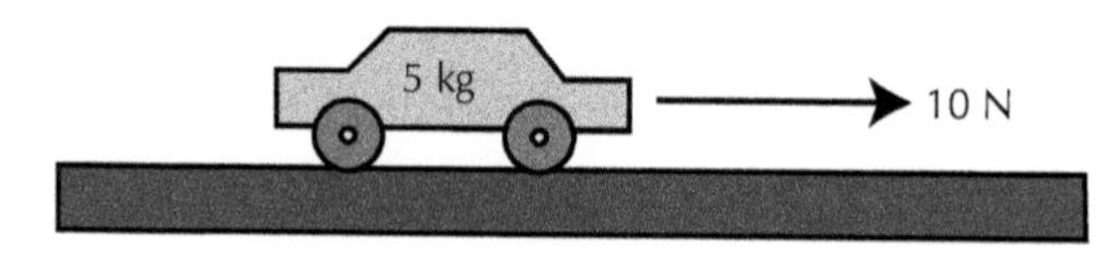

The mass of the car is 5 kg. The car is travelling at a constant speed.

The force of friction acting on the car is

A 0 N

B 2 N

C 5 N

D 10 N

E 50 N.

N5 Physics

Practice Papers for SQA Exams

Physics Section 2

Fill in these boxes:

Name of centre

Town

Forename(s)

Surname

Try to answer all of the questions in the time allowed.

Total marks — 110

Section 1 — 20 marks

Section 2 — 90 marks

Read all questions carefully before attempting.

You have 2 hours to complete this paper.

Write your answers in the spaces provided, including all of your working.

MARKS
Do not write in this margin

SECTION 2

1. A skateboarder is practising on a ramp. The mass of the skateboarder and the board is 70 kg.

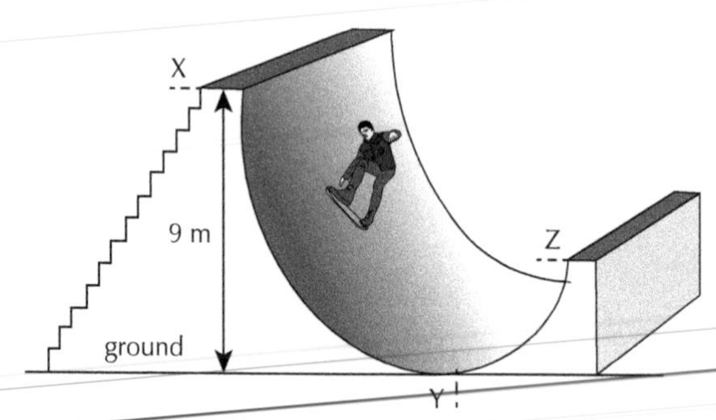

(a) Calculate the increase in potential energy of skateboarder and board in moving from the ground to position X. **3**

Space for working and answer

(b) The skateboarder moves from X to Z.

(i) At what point on the ramp is the kinetic energy of the skateboarder greatest? **1**

(ii) The vertical speed of the skateboarder at Z is $5{\cdot}5\ ms^{-1}$.
Calculate the maximum height above Z the skateboarder can rise to. **5**

Total marks 9

MARKS Do not write in this margin

2. An electric kettle is filled with water.

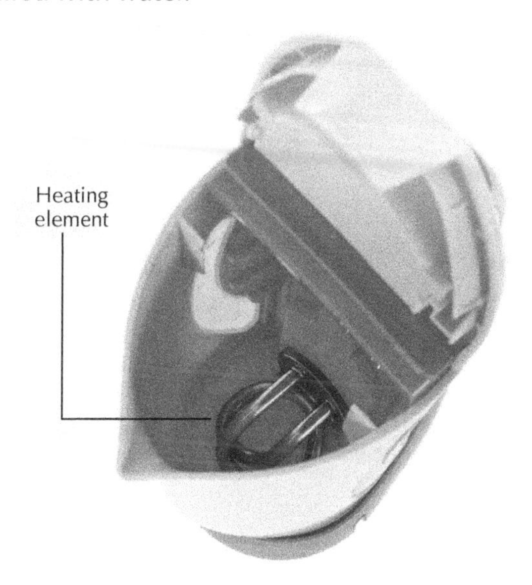

The heating element in the kettle is used to raise the temperature of the water.
The heating element is rated at 240 V, 2 kW.

(a) The heating element is switched on for 40 s.

Calculate the electrical energy supplied to the heating element.

Space for working and answer **3**

(b) The mass of water inside the kettle is 0·3 kg. The temperature of the water rises from 25 °C to 84 °C during the time the heating element is switched on.

Calculate the heat energy gained by the water.

Space for working and answer **4**

(c) Explain why the heat energy gained by the water is less than the electrical energy supplied to the heating element. **2**

Total marks 9

MARKS
Do not write in this margin

3. An LDR is used as a light sensor in the circuit below. When the light level falls below a certain value, the output causes a switching circuit to turn on a street light.

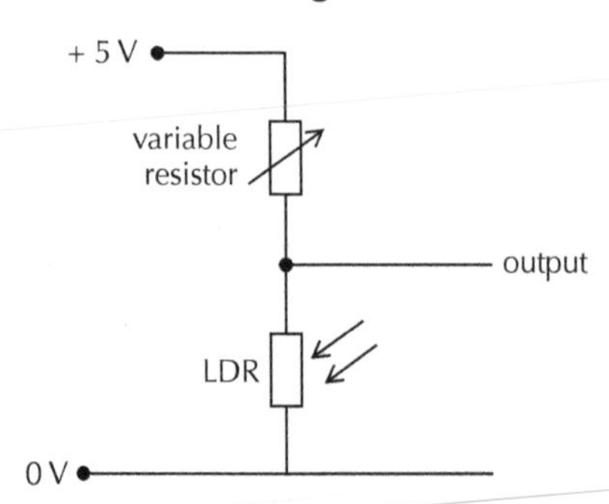

(a) When the voltage across the LDR is 0·7 V, the circuit causes the street light to switch on.

The resistance of the variable resistor is set at 8600 Ω.

Calculate the resistance of the LDR when the voltage across the LDR is 0·7 V. **4**

Space for working and answer

(b) The graph shows how the resistance of the LDR changes with temperature.

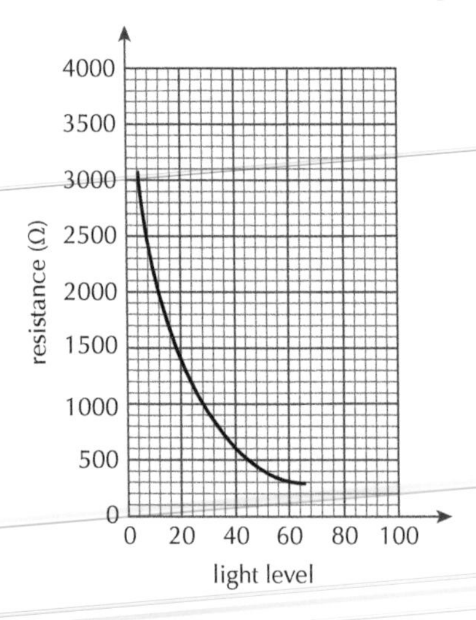

(i) State what happens to the resistance of the LDR as the light level increases. **1**

(ii) Use the graph to determine the light level at which the street light is switched on. **1**

(c) The switching circuit is connected as shown. When there is a current in the relay coil, the relay switch closes.

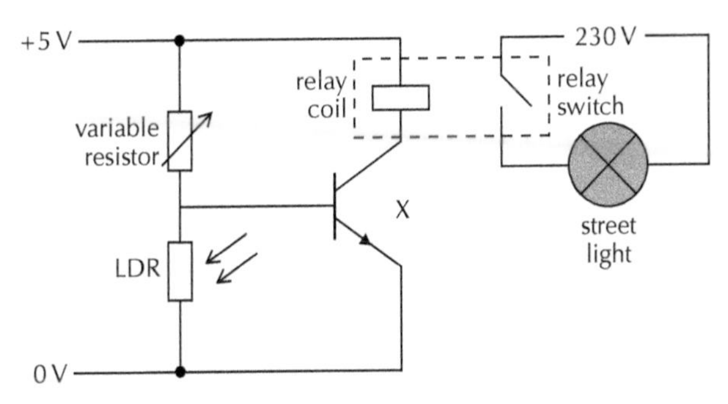

(i) Name component X. **1**

(ii) Explain how the circuit operates to switch on the street light. **3**

Total marks 10

MARKS
Do not write in this margin

4. The photoelectric effect was discovered towards the end of the 19th century. Experiments showed that electrons could be emitted from the surface of a metal by illuminating the metal with light.

The electrons are only emitted from the surface of the metal if the light is above a certain frequency. This is known as the threshold frequency.

The energy needed to remove an electron from the surface of a metal is called the work function. The work function, in Joules, of a metal is given by the equation:

$$E = hf_0$$

where h is Planck's constant ($h = 6{\cdot}63 \times 10^{-34}$ J s) and f_0 is the threshold frequency.

(a) The table below gives the work function of different metals.

Metal	***Work function* (J)**
Zinc	$6{\cdot}9 \times 10^{-19}$
Gold	$7{\cdot}8 \times 10^{-19}$
Sodium	$3{\cdot}6 \times 10^{-19}$
Potassium	$3{\cdot}2 \times 10^{-19}$

An unknown metal, metal X, is found to have a threshold frequency of $1{\cdot}18 \times 10^{15}$ Hz. Identify metal X.

Space for working and answer **3**

(b) Calculate the longest wavelength of light that would allow an electron to be emitted from the metal potassium. **3**

Total marks 6

5. When white light is passed through a glass prism, a spectrum of colours is produced.

A student makes the statement:

'The colours are produced by the glass prism.'

Use your knowledge of physics to explain why the student is incorrect. **3**

MARKS Do not write in this margin

6. A pulsar is a rapidly spinning star that emits radio waves.

The pulsar PSR B1257 + 12 is $9{\cdot}46 \times 10^{18}$ m from the Earth.

The radio waves are received by a telescope on Earth.

(a) What is the speed of radio waves? **1**

(b) Calculate the time taken for the radio waves to reach Earth.

Space for working and answer **3**

(c) The period of rotation of PSR B1257 + 12 is 6·2 ms.
Calculate the number of rotations in 1 second.

Space for working and answer **3**

(d) As they grow older, all pulsars slow down.

Astronomers measuring another pulsar in the Crab Nebula have created an equation to predict the spin rate of the pulsar in the future.

$P = 0{\cdot}033 + 0{\cdot}000013T$

where P is the period of rotation of the Crab Nebula pulsar in seconds and T is the number of years since today.

Calculate the period of rotation of the Crab Nebula Pulsar 10 000 years in the future.

Space for working and answer **2**

Total marks 9

7. Infrared radiation is used to send information via fibre optic cables.

(a) The diagram shows an infrared ray incident on a glass fibre.

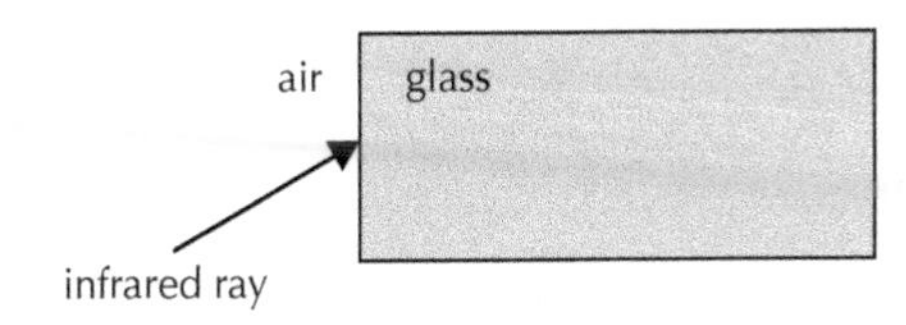

Complete the diagram to show the path of the infrared ray as it enters the glass.

Indicate on your diagram the normal, the angle of incidence and the angle of refraction. **2**

(b) Infrared radiation is part of the electromagnetic spectrum.

A section of the electromagnetic spectrum is shown in the diagram below.

Infrared radiation	Visible light	Ultraviolet light	**A**	Gamma radiation

(i) Identify radiation A. **1**

(ii) Which of these electromagnetic waves has the greatest energy? **1**

(iii) State **one** property that all electromagnetic waves have in common. **1**

Total marks 5

MARKS
Do not write in this margin

8. (a) Rutherfordine is a mineral that contains Uranium.

The activity of 1·0 kg of pure Rutherfordine is $1{\cdot}3 \times 10^8$ decays per second.

A sample of mass 0·4 kg contains 50% Rutherfordine. The other 50% is not radioactive.

Calculate the activity of the sample in becquerels.

Space for working and answer **4**

(b) The equivalent dose limit for a worker is 0·2 mSv per day.

The worker is in contact with the radiations for 8 hours each day.

The table below gives information on the radiation exposure.

Type of radiation	***Absorbed dose per hour***
Gamma	0·1 μSv
Fast neutrons	1 μSv
Slow neutrons	4 μSv

Show, by calculation, whether the equivalent dose limit is exceeded.

Space for working and answer **5**

Total marks 9

9. A volleyball player strikes a ball. The ball leaves the hand horizontally at 12 m s^{-1}. It hits the ground at a distance of 5·8 m from the point where it was struck.

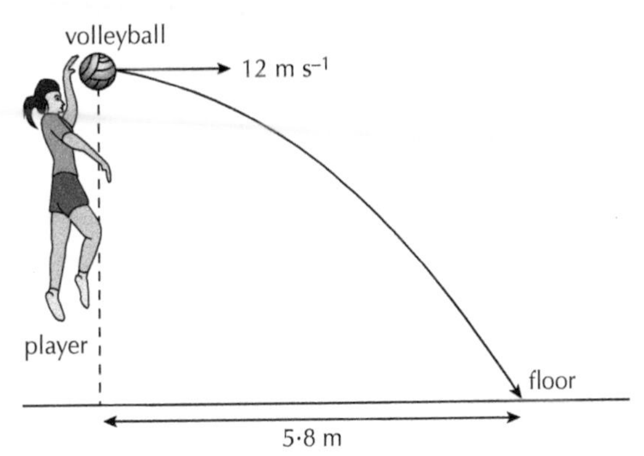

Assume that air resistance is negligible.

(a) Explain why the shape of the path taken by the ball is curved. **2**

(b) Calculate the time of flight of the ball. **3**

Space for working and answer

(c) Calculate the vertical speed of the ball as it reaches the ground. **3**

Space for working and answer

(d) Sketch a graph of vertical speed against time for the ball. Numerical values are required on both axes. **2**

Total marks 10

MARKS
Do not write in this margin

10. A spacecraft is on the surface of the Moon.

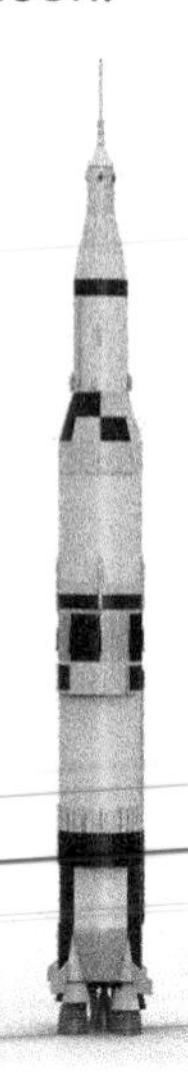

The spacecraft has a total mass of $2{\cdot}8 \times 10^6$ kg.

The spacecraft's engines produce a total force of $3{\cdot}5 \times 10^7$ N.

(a) (i) Calculate the weight of the spacecraft on the surface of the Moon. **3**

Space for working and answer

(ii) Sketch a diagram showing the forces acting on the spacecraft immediately after take-off from the Moon. You must name the forces and show their direction. **2**

(iii) Calculate the acceleration of the spacecraft as it takes off from the Moon. **4**

Space for working and answer

(b) An identical spacecraft is launched from the surface of the Earth. The mass of the spacecraft and the engine force are the same as before.

Is the acceleration of the spacecraft as it takes off from the Earth greater than, less than, or equal to the acceleration as it takes off from the Moon?
You must justify your answer by calculation.

Space for working and answer **4**

Total marks 13

MARKS
Do not write in this margin

11.

Fraunhofer lines

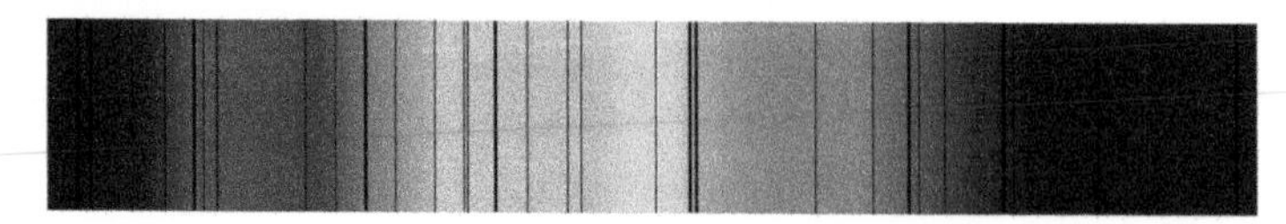

The emission spectrum of the Sun has a number of dark lines in it. The dark lines are caused by absorption by cooler gases just above the hot visible surface that we see. The lines are called Fraunhofer lines after Josef von Fraunhofer (Bavaria, 1787–1826), who discovered the dark lines. He developed a device called a spectroscope to view the absorption lines. The absorption lines indicate the elements that are present at the Sun's surface.

The intensity of the absorption lines for an element can tell us how much of the element is present: the more of the element that is at the surface, the more absorption takes place and the darker the line. Such measurements show that the Sun's atmosphere consists of 72% hydrogen, 26% helium and 2% heavier elements.

(a) Name the device used by Fraunhofer in the discovery of absorption lines. **1**

(b) Explain how the absorption lines are used to determine how much of each element is present. **1**

(c) A line spectrum from the Sun is shown below along with the line spectra of the elements sodium, helium, hydrogen and nitrogen.

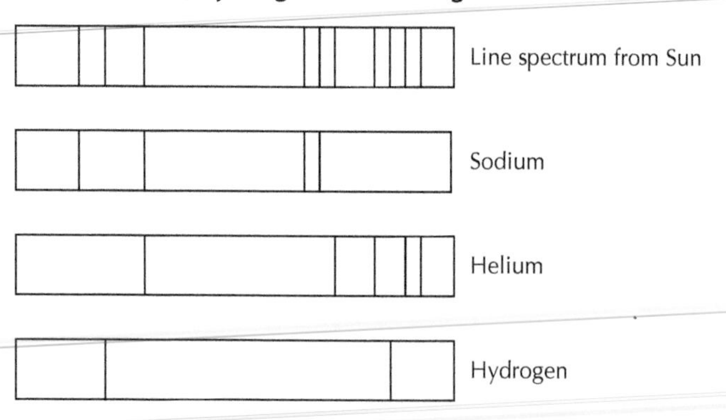

Identify the elements present in the Sun. **2**

Total marks 4

12. A news report states:

'We're probably being too optimistic about finding life elsewhere in the Universe – it's very unlikely.'

Using your knowledge of physics, comment on the above statement. **3**

[END OF QUESTION PAPER]

Practice Exam C

N5 Physics

Practice Papers for SQA Exams

Physics Section 1

Fill in these boxes:

Name of centre

Town

Forename(s)

Surname

Try to answer all of the questions in the time allowed.

Total marks — 110

Section 1 — 20 marks

Section 2 — 90 marks

Read all questions carefully before attempting.

You have 2 hours to complete this paper.

Write your answers in the spaces provided, including all of your working.

SECTION 1

Objective Test

1. A curling stone is pushed across ice towards a target.

The mass of the curling stone is 18 kg. The curler's hand stays in contact with the stone for a distance of 2·5 m. The stone is pushed with an average force of 15 N.

The maximum kinetic energy of the curling stone is

A 1·2 J

B 6·0 J

C 37·5 J

D 45·0 J

E 270 J.

2. Current can be defined as

A the rate of charge flowing past a point

B the rate of energy transferred

C the rate of energy transformed

D the rate of change of energy

E the rate of change of charge.

3. The specific heat capacity of a substance is the energy required to

A change the state of 1 kg of a substance

B change the temperature of the substance without changing its state

C change the temperature of 1 kg of a substance by 1 °C

D evaporate 1 kg of a substance

E melt 1 kg of a substance.

4. Identify the following circuit symbol.

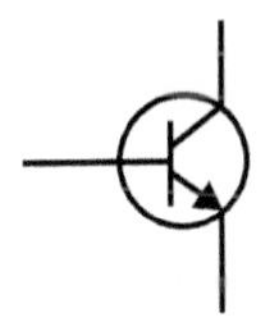

A LED

B NPN transistor

C N-channel enhancement MOSFET

D Thermistor

E Capacitor

5. A circuit is set up as shown.

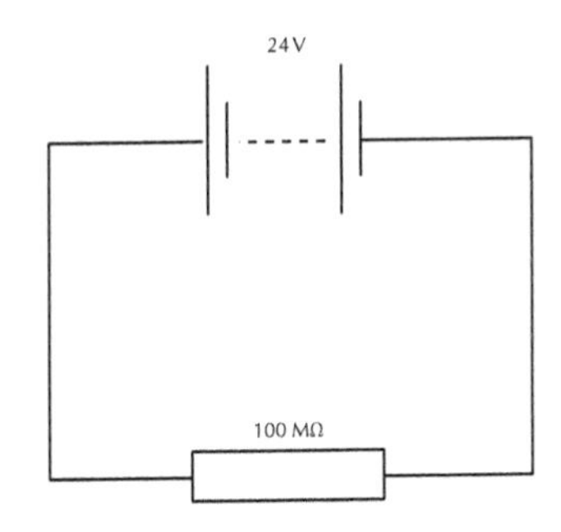

The power supplied to the resistor is

A 0·24 W

B 5·76 W

C $5{\cdot}76 \times 10^{-6}$ W

D $2{\cdot}4 \times 10^{-7}$ W

E 416·7 W.

6. In the diagrams below, each resistor has a resistance of 2 Ω.

Select the combination which has the **smallest** total resistance between terminals P and Q.

A 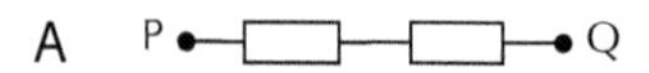

B 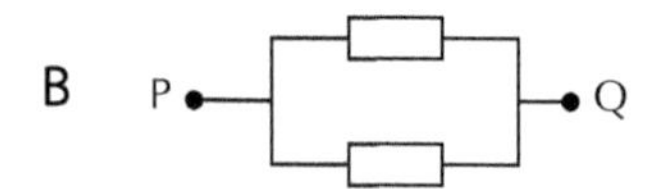

C 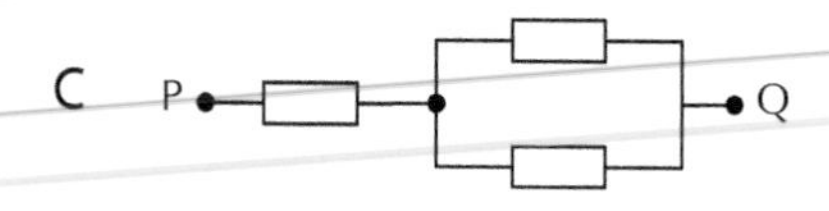

D 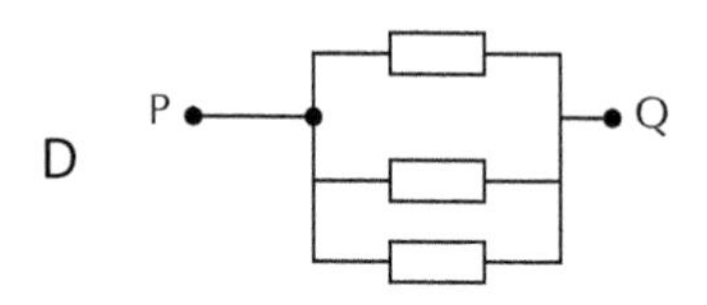

E 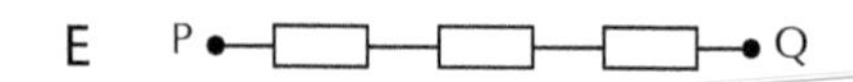

7. An elephant has a mass of 1800 kg. The elephant stands on four feet, each with contact area 0·5 m^2. The pressure exerted by the four feet on the surface of the Earth is

A 900 Pa

B 3600 Pa

C 8820 Pa

D 17 640 Pa

E 35 280 Pa.

8. A student writes the following statements in their physics notebook.

I The pressure of a fixed mass of gas varies directly as its Kelvin temperature, provided the volume of the gas remains constant.

II The pressure of a fixed mass of gas varies directly as its volume, provided the temperature of the gas remains constant.

III The volume of a fixed mass of gas varies directly as its Kelvin temperature, provided the pressure of the gas remains constant.

Which of the statements is/are correct?

A I only

B I and II only

C I and III only

D II and III only

E I, II and III

9. A sound engineer sets up the apparatus shown to measure the speed of sound in air. The time taken for the sound to travel between the two microphones is recorded on the electronic timer.

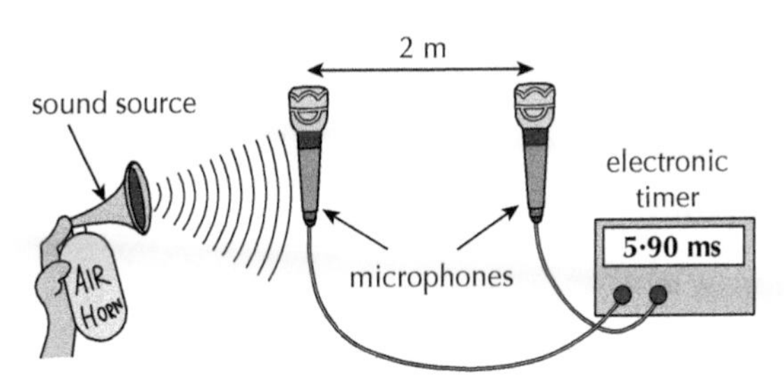

The calculated speed of sound in air is

A 0·34 m s^{-1}

B 11·8 m s^{-1}

C 295 m s^{-1}

D 330 m s^{-1}

E 339 m s^{-1}.

10. Sound is a longitudinal wave. When a sound wave travels through air, the particles of air

A vibrate at 45° to the direction of the wave

B vibrate at 90° to the direction of the wave

C vibrate in the same direction as the direction of the wave

D vibrate in the opposite direction to the direction of the wave

E vibrate in all directions.

11. Which row shows the members of the electromagnetic spectrum in order of increasing frequency?

A	visible light	ultraviolet	infrared
B	visible light	infrared	ultraviolet
C	visible light	X-rays	gamma rays
D	visible light	X-rays	ultraviolet
E	visible light	ultraviolet	X-rays

12. For a ray of light travelling from plastic into air, which of the following statements is/are correct?

I The speed of light increases.

II The angle of refraction is greater than the angle of incidence.

III The angle of refraction is less than the angle of incidence.

A I only

B II only

C III only

D I and II only

E I and III only

13. Light travels from air to glass as shown.

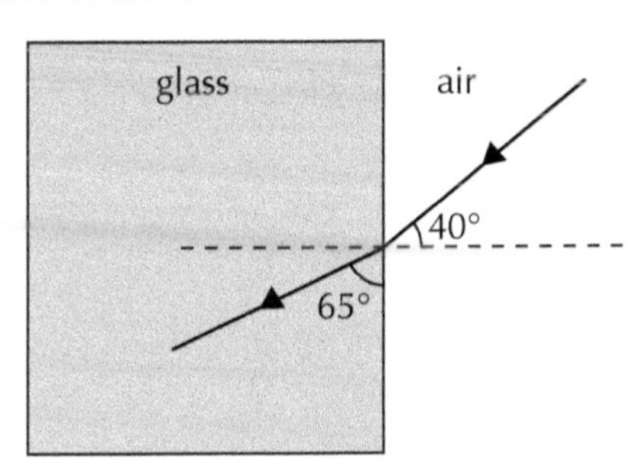

The angle of refraction is

A 25°

B 40°

C 50°

D 65°

E 105°.

14. Which row correctly describes alpha, beta and gamma radiation?

	Alpha	***Beta***	***Gamma***
A	fast-moving electron	helium nucleus	electromagnetic radiation
B	electromagnetic radiation	helium nucleus	fast-moving electron
C	helium nucleus	fast-moving electron	electromagnetic radiation
D	helium nucleus	electromagnetic radiation	fast-moving electron
E	electromagnetic radiation	fast-moving electron	helium nucleus

15. One gray is equal to

A one joule per second

B one becquerel per kilogram

C one sievert per second

D one joule per kilogram

E one sievert per second.

16. Which row contains two vector quantities and one scalar quantity?

A	displacement	velocity	acceleration
B	displacement	velocity	speed
C	distance	velocity	force
D	distance	displacement	force
E	speed	acceleration	time

17. A car starting from rest accelerates at 6 m s^{-2}. The speed of the car after 3 s is

A 2 m s^{-1}

B 3 m s^{-1}

C 6 m s^{-1}

D 18 m s^{-1}

E 36 m s^{-1}.

18. The diagram shows the horizontal forces acting on a box.

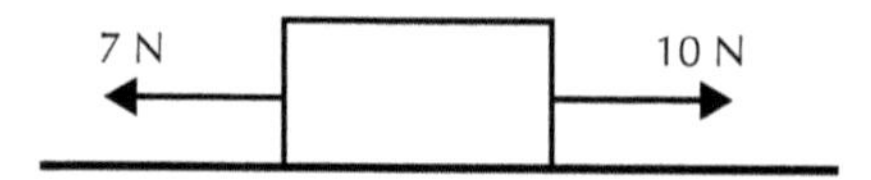

The box accelerates at 2·0 m s^{-2}.

The mass of the box is

A 1·5 kg

B 3·5 kg

C 5·0 kg

D 6·0 kg

E 20 kg.

19. A ball rolls along a table and leaves it at point A.

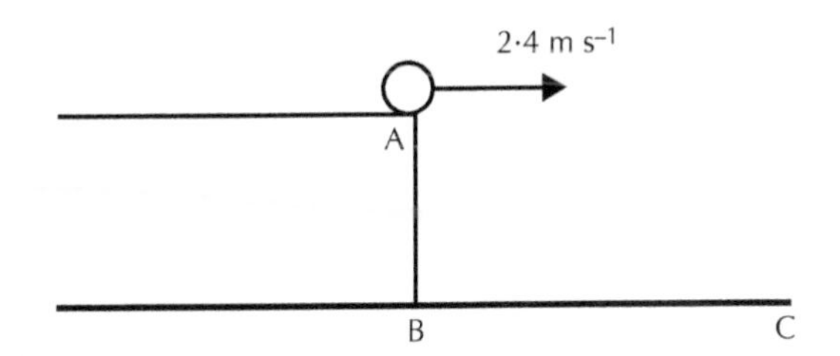

The horizontal speed of the ball at A is 2·4 m s^{-1}.

The ball takes 0·75 s to travel from A to C.

The distance BC is

A 0·3 m

B 1·8 m

C 3·2 m

D 7·4 m

E 18 m.

20. A person is walking in the woods. The person's feet exert a downward force on the ground.

Which of the following completes the 'Newton pair' of forces?

A The force of the ground on the person's feet.

B The force of the person's feet on the ground.

C The force of gravity on the person's feet.

D The force of the Earth on the person.

E The force of gravity on the person.

N5 Physics

Practice Papers for SQA Exams

Physics Section 2

Fill in these boxes:

Name of centre

Town

Forename(s)

Surname

Try to answer all of the questions in the time allowed.

Total marks — 110

Section 1 — 20 marks

Section 2 — 90 marks

Read all questions carefully before attempting.

You have 2 hours to complete this paper.

Write your answers in the spaces provided, including all of your working.

SECTION 2

1. (a) A student sets up the following circuit.

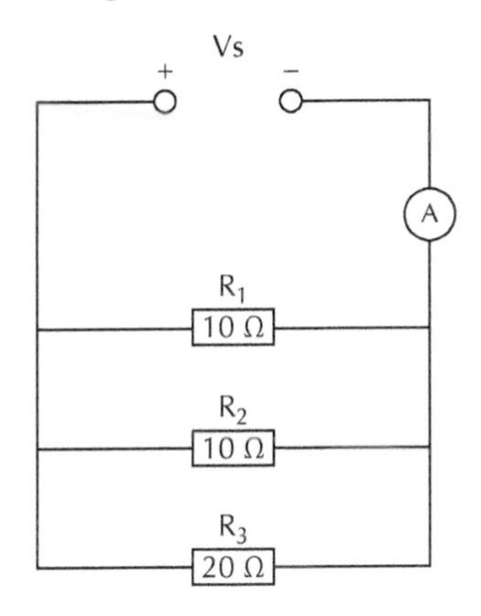

(i) Calculate the total resistance of the circuit. **3**

Space for working and answer

(ii) The reading on the ammeter is 0·5 A.

Calculate the supply voltage V_S. **3**

Space for working and answer

(b) Calculate the quantity of charge that will flow through resistor R_3 in 5 minutes. **5**

Space for working and answer

Total marks 11

MARKS Do not write in this margin

2. A meteorite of mass 150 kg enters the Earth's atmosphere with a speed of 2200 m s^{-1}. As the result of friction between the atmosphere and the meteorite, the speed of the meteorite decreases to 800 m s^{-1}.

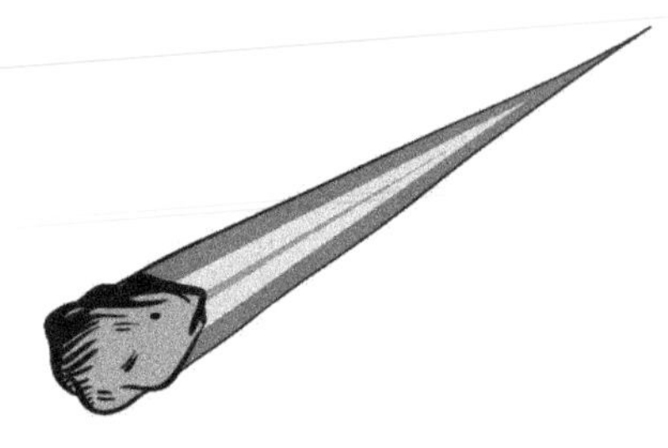

The material from which the meteorite is made has a specific heat capacity of 1050 J kg^{-1} °C.

(a) Calculate the rise in temperature of the meteorite on entering the Earth's atmosphere.

Space for working and answer **6**

(b) Is the actual temperature change of the meteorite greater than, the same as, or less than the value calculated in part (a)?

You must explain your answer. **2**

Total marks 8

MARKS
Do not write in this margin

3. The apparatus shown in the diagram is used to investigate the relationship between the temperature and pressure of a fixed mass of air, which is kept at a constant volume.

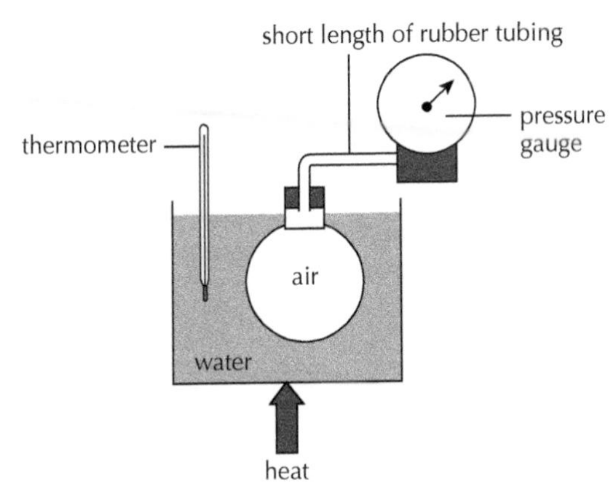

(a) The pressure of the air is 121 kPa when its temperature is 17 °C.

The temperature of the air rises to 69 °C.

Calculate the new pressure of the air inside the flask. **4**

Space for working and answer

(b) Explain, in terms of the movement of gas molecules, what happens to the pressure of the air as its temperature is increased. **2**

(c) Suggest **one** improvement that could be made to the apparatus to give a more accurate result. You must explain clearly why your suggestion improves the design of the apparatus. **2**

Total marks 8

MARKS
Do not write in this margin

4. A diffraction grating is made up of a series of narrow, parallel slits. When red light shines on the diffraction grating a set of bright dots appear on the screen. These bright areas are known as maxima.

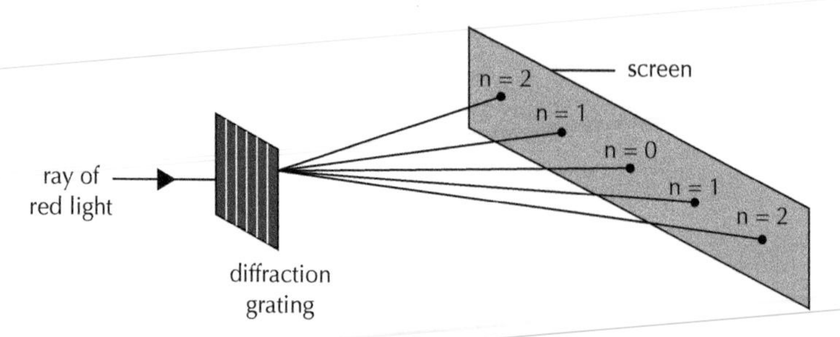

The first maximum is at $n = 0$. This is called the zero order maximum. At $n = 1$ is the first order maximum and at $n = 2$, the second order maximum.

For a maximum to occur:

$$n\lambda = d \sin \theta$$

where

n is the order of the maximum

λ is the wavelength of the light in metres

d is the slit separation in metres

θ is the angle from the zero order maximum.

(a) The red light has a wavelength of 700 nm. The slit separation on the diffraction grating is $1{\cdot}6 \times 10^{-6}$ m.

Calculate the angle between the zero order maximum and the first order maximum. **3**

Space for working and answer

(b) The red light is replaced with a blue light source. The blue light has a wavelength of 450 nm.

(i) Calculate the angle between the zero order maximum and the first order maximum for blue light. **3**

Space for working and answer

(ii) Using your answers to part (a) and part (b) (i), comment on the appearance of the maxima on the screen for red light compared to the maxima on the screen for blue light. **2**

Total marks 8

MARKS
Do not write in this margin

5. A student observes a pencil in a glass of water. The pencil appears to change shape at the surface of the water. When the student removes the pencil from the water it is perfectly straight.

Using your knowledge of physics, comment on why the pencil looks visually different when in the water. **3**

6. A spring is used to demonstrate wave motion.

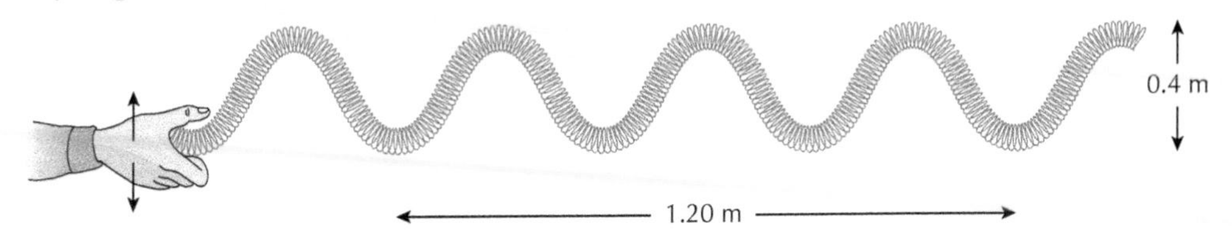

(a) (i) What is the amplitude of the wave? **1**

(ii) Calculate the wavelength of the wave. **1**

(b) There are 45 waves produced on the spring each minute.

(i) Calculate the speed of the waves.

Space for working and answer **4**

(ii) Calculate the period of the waves.

Space for working and answer **3**

Total marks 9

MARKS
Do not write in this margin

7. The diagram shows a communications satellite orbiting the Earth.

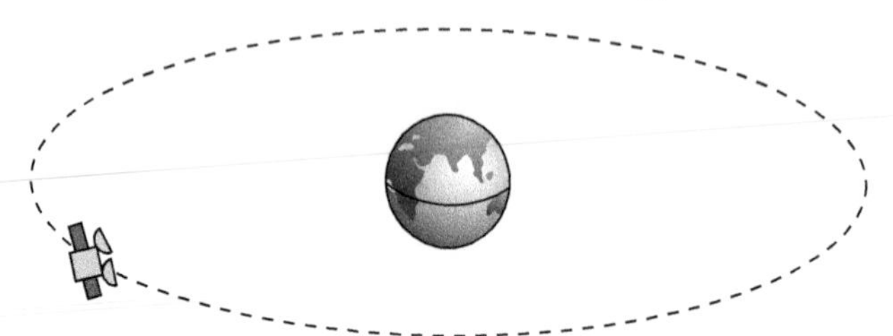

The communications satellite orbits at a height of approximately 36 000 km above the Earth. The satellite uses microwaves to send information to Earth.

(a) (i) What is the speed of microwaves? **1**

(ii) Calculate the time taken for the microwaves to travel from the satellite to the Earth's surface.

Space for working and answer **3**

(b) Explain why the satellite stays in orbit around the Earth. **2**

Total marks 6

8. The following table contains information about two radioactive sources.

Radioactive source	***Activity* (MBq)**	***Half-life* (days)**
X	850	18
Y	1800	6

(a) Calculate the number of decays of source X in 1 minute 30 s. **3**

Space for working and answer

(b) The radioactive sources can be disposed of when their activity reaches 30 MBq.

Show by calculation which source, X or Y, will be first to reach an activity of 30 MBq. **5**

Space for working and answer

Total marks 8

9. In an orienteering event, competitors navigate from the start to a series of control points around a set course. At each control point, the competitors receive a three-digit code that they must enter into an electronic unit they carry with them. The first competitor to show evidence of visiting all the control points wins the event.

Two orienteerers, Sarah and Jennifer, take part in a race in a flat area.

From the start point, Sarah runs 800 m north (000), then 700 m west (270) to arrive at the first control point. It takes her 12 minutes to reach the first control point.

(a) State the difference between vector and scalar quantities. **2**

(b) (i) By scale drawing, or otherwise, find the displacement of Sarah from the start point.

Space for working and answer **4**

(ii) Calculate the average velocity of Sarah between the start and the first control point.

Space for working and answer

3

(c) Jennifer leaves the start point at the same time as Sarah.

Jennifer follows exactly the same route at an average speed of $1{\cdot}9\ \text{m s}^{-1}$.

Show by calculation who arrives first at the control point.

Space for working and answer

4

Total marks 13

10. A cyclist is competing in a road race.

The graph below shows how the cyclist's speed changes with time during the race.

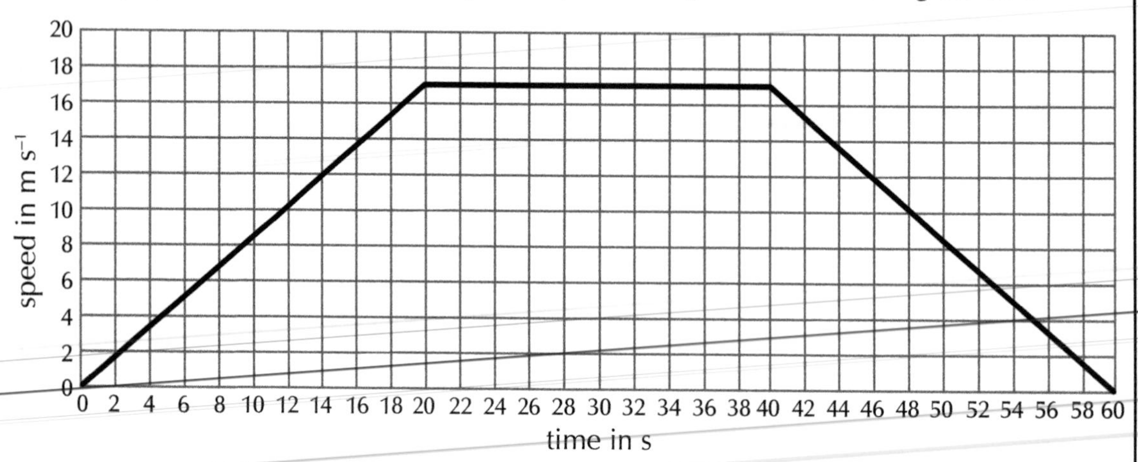

(a) Calculate the initial acceleration of the cyclist. **3**

Space for working and answer

(b) Calculate the distance travelled by the cyclist whilst decelerating. **3**

Space for working and answer

MARKS
Do not write in this margin

(c) The diagram shows the cyclist during the race.

(i) Suggest one way in which the cyclist **reduces** friction. **1**

(ii) Suggest one place where the cyclist **requires** friction. **1**

Total marks 8

11.

Black holes

When a star reaches the end of its life, it has no fuel left to continue nuclear fusion. At a certain stage the gravitational forces overcome the thermal forces produced by nuclear fusion and the star collapses very rapidly. The compression is so great that the outer layers of material fall inwards then rebound, setting up massive shock waves, which result in a huge explosion called a supernova.

Black holes will only be formed if the mass left after the supernova is sufficiently large. Our Sun is much too small, with a mass of 2×10^{30} kg, which is commonly referred to as 1 solar mass. If the core of a star after a supernova is greater than 3 solar masses, then a black hole will be produced. The gravitational field around a black hole is so high that no radiation, not even light, can escape.

(a) Calculate the smallest star mass after a supernova that would form a black hole. **2**

Space for working and answer

(b) The escape velocity from a black hole is defined as the minimum velocity required to escape the black hole's gravitational pull.

The escape velocity from the surface of an object of mass M and radius R is given by the formula

$$v = \sqrt{\frac{2GM}{R}}$$

where G is the gravitational constant ($G = 6{\cdot}7 \times 10^{-11}$ N m^2 kg^{-2}).

Calculate the escape velocity of a black hole of mass $2{\cdot}04 \times 10^{31}$ kg and radius 30 km. **3**

Space for working and answer

Total marks 5

MARKS
Do not write in this margin

12. During the design of modern cars, they are tested in wind tunnels. This allows engineers to see how air flows over the body of the car.

Using your knowledge of physics, comment on why wind tunnel testing is useful in car design. **3**

[END OF QUESTION PAPER]

ANSWERS

Answers to Practice Exams

Practice Exam A

Section 1

Question	Response	Mark	Top Tips
1.	C	1	Don't forget to square the speed in the equation for kinetic energy.
2.	C	1	Use the values of P and R to calculate the current, I, first before using an equation that involves current, charge and time.
3.	D	1	Remember that the definition of one volt is one joule per coulomb. In other words, energy per unit of charge.
4.	C	1	First calculate the total resistance of the parallel branch. The total resistance is the sum of this value and the 10 Ω resistor.
5.	B	1	The current is the same at all points in a series circuit whereas the voltage is split up across the components.
6.	D	1	The Δ symbol in the specific heat capacity equation is the Greek letter delta. ΔT or delta T means a change in temperature.
7.	C	1	The graph is showing a constant volume. What is the relationship between pressure and temperature for a fixed volume of gas?
8.	A	1	To convert from degrees Celsius to Kelvin, add 273 to each number. Remember 0 K = –273 °C.
9.	D	1	The speed of all signals in the electromagnetic spectrum is 3×10^8 m s^{-1}. This value can be found in the data sheet as the 'speed of light in air'.
10.	B	1	Frequency needs to be calculated first. Remember that frequency is the number of waves per second.

Question	Response	Mark	Top Tips
11.	B	1	Try making up a mnemonic with the letters R, M, I, V, U, X, G to help remember the correct order of the signals in the electromagnetic spectrum.
12.	E	1	When light moves from glass into air, it always speeds up as air is less dense than glass. The frequency remains the same, therefore the wavelength must increase also.
13.	C	1	When moving from air to glass, light bends towards the normal. When moving from glass to air, light bends away from the normal.
14.	D	1	Think about what each of the radiations are made of – which one would be most likely to interact with other atoms?
15.	E	1	The half-life is the time taken for the activity to fall by half. In this case the activity falls by half every 8 hours.
16.	C	1	Vector quantities have size **and** direction. Think about which quantities it makes sense to give a direction to.
17.	C	1	To calculate the instantaneous speed at the light gate, you need to know the time it takes to pass through the light gate.
18.	A	1	An object with constant acceleration changes its velocity by the same amount each second. An object with a constant velocity has a velocity that does not change.
19.	D	1	Remember you can find the gravitational field strength of the Moon in the data sheet at the front of the exam paper.
20.	C	1	If an object A exerts a force on object B, then object B will exert an equal but opposite force on object A.

Exam A

Section 2

Question			Expected Response		Max mark	Top Tips
1.	(a)	(i)	$R_t = R_2 + R_3$ $= (4 + 8)$ $= 12\ (\Omega)$ $I = \frac{V}{R}$ $= \frac{12}{12}$ $= 1\ \text{A}$	 (1) (1) (1) (1)	4	Current cannot flow across a gap in a circuit so when switch S is open, no current will flow in resistor R_1. The circuit can be treated as two resistors in series. Remember that the current in a series circuit is the same at all points in the circuit.
	(a)	(ii)	$\frac{1}{R_t} = \frac{1}{R_2} + \frac{1}{R_3}$ $\frac{1}{R_t} = \frac{1}{8} + \frac{1}{4}$ $\frac{1}{R_t} = \frac{1}{8} + \frac{2}{8}$ $R_t = \frac{8}{3}$ $R_t = 2{\cdot}67\ (\Omega)$ $V_2 = \left(\frac{R_2}{R_1 + R_2}\right) V_s$ $= \left(\frac{2{\cdot}67}{10{\cdot}67}\right) 12$ $= 3\ \text{V}$	(1) (1) (1) (1) (1)	5	Because switch S is closed, this makes things a little more difficult. To make the problem easier to solve, the parallel section of the circuit can be simplified into one resistor. The circuit can then be treated like two resistors in series as with (a) (i). The potential difference across R_2 can also be found by using $V = IR$.

Question			Expected Response	Max mark	Top Tips
	(b)		$V_{R_3} = 16\ \text{V}$ $I = \frac{V}{R}$ $I = \frac{16}{8}$ $I = 2\ \text{A}$ (1) Lamp: $P = IV$ (1) $= 2 \times 6$ $= 12\ (\text{W})$ (1) Resistor: $P = IV$ $= 2 \times 16$ $= 32\ (\text{W})$ (1) Greater power dissipated in resistor R_3. (1)	5	Remember in a series circuit that the voltage from the supply is split up across the components in the circuit. If the lamp is operating at 6 V then the remaining voltage from the supply must be across resistor R_3. This allows the current to be calculated, as the current is the same at all points in a series circuit. The power dissipated in each component can then be calculated.
2.	(a)	(i)	$T_1 = 6 + 273 = 279\ \text{K}$ $T_2 = 21 + 273 = 294\ \text{K}$ (1) $\frac{P_1}{T_1} = \frac{P_2}{T_2}$ (1) $\frac{2{\cdot}36 \times 10^6}{279} = \frac{P_2}{294}$ (1) $P_2 = 2{\cdot}49 \times 10^6\ \text{Pa}$ (1)	4	When performing gas laws calculations, temperature must always be converted to Kelvin. None of the gas laws relationships involving temperature will work when using degrees Celsius. $0\ ^\circ\text{C} = 273\ \text{K}$ $0\ \text{K} = -273\ ^\circ\text{C}$
	(a)	(ii)	More molecules/atoms/particles hit container walls per second. (1) Collisions with walls harder/larger force. (1)	2	The kinetic model describes pressure in a container as being caused by the collision of particles with the container walls. An increase in temperature causes those particles to move around faster.
	(b)		Pressure continues to increase as temperature increases. (1) The fire extinguisher has a maximum pressure it can maintain. (1)	2	Problem solving question. Every container will have a maximum pressure it can withstand before the gas inside is released.

Question			Expected Response	Max mark	Top Tips
3.			Demonstrates no understanding, 0 marks. Demonstrates limited understanding, 1 mark. Demonstrates reasonable understanding, 2 marks. Demonstrates good understanding, 3 marks. **1 mark:** The student has demonstrated a limited understanding of the physics involved. The student has made some statement (s) which is/are relevant to the situation, showing that at least a little of the physics within the problem is understood. **2 marks:** The student has demonstrated a reasonable understanding of the physics involved. The student makes some statement (s) which is/are relevant to the situation, showing that the problem is understood. **3 marks:** The maximum available mark would be awarded to a student who has demonstrated a good understanding of the physics involved. The student shows a good comprehension of the physics of the situation and has provided a logically correct answer to the question posed. This type of response might include a statement of the principles involved, a relationship or an equation, and the application of these to respond to the problem. This does not mean the answer has to be what might be termed an 'excellent' answer or a 'complete' one.	**3**	Open-ended question. Try and write down everything you know about the physics described in the question and, if possible, include a physics law or equation.

Question			Expected Response	Max mark	Top Tips
4.	(a)		Solar cell A: Percentage efficiency $= \frac{0{\cdot}7 \times 3{\cdot}8}{10} \times 100$ (1) $= 26{\cdot}6\%$ (1) Solar cell B: Percentage efficiency $= \frac{0{\cdot}6 \times 3{\cdot}5}{10} \times 100$ (1) $= 21\%$ (1) Solar cell C: Percentage efficiency $= \frac{0{\cdot}98 \times 2{\cdot}3}{15} \times 100$ (1) $= 15\%$ (1)	**6**	The National 5 exam will include unfamiliar equations. The question will explain clearly the quantities given in the equation so read the question carefully. When performing a series of calculations, as in this question, it is good practice to label each individual calculation (e.g. Solar cell A). This way, the marker can clearly see which solar cell has the highest percentage efficiency.
	(b)		$f = \frac{v}{\lambda}$ (1) $f = \frac{3 \times 10^{8}}{545 \times 10^{-9}}$ (1) $f = 5{\cdot}5 \times 10^{14}\ \text{Hz}$ (1)	**3**	Remember that all signals within the electromagnetic spectrum travel at the speed of light $3 \times 10^{8}\ \text{m s}^{-1}$. This value is given in the data sheet at the front of the exam paper. Make sure you are familiar with the information given in the data sheet.
5.	(a)		Energy	**1**	
	(b)	(i)	Transverse	**1**	All waves from the electromagnetic spectrum are transverse waves. Sound is a longitudinal wave.

Question			Expected Response	Max mark	Top Tips
	(b)	(ii)	A transverse wave is one in which the particles vibrate at right angles to the direction of the wave. (1) A longitudinal wave is one in which the particles vibrate parallel to the direction of the wave. (1)	**2**	When describing differences make sure that you clearly describe what is different about **both** items mentioned. For example, if a question asked for the differences between vector and scalar quantities, an answer of 'vector quantities have direction' would get 0 marks. The answer would have to be 'scalar quantities have magnitude only, vector quantities have magnitude and direction'.
	(c)		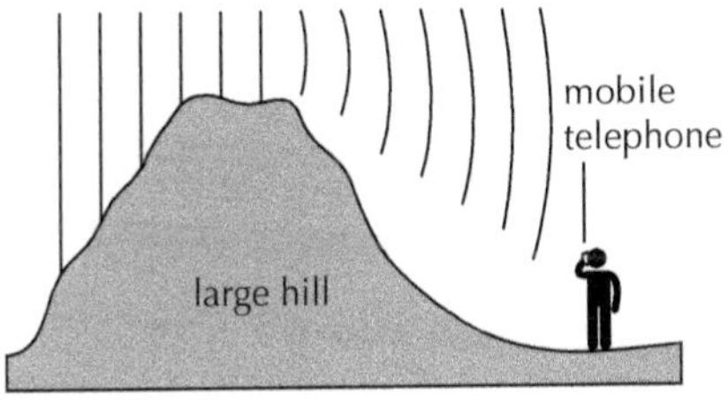 mobile telephone large hill Correct frequency throughout. (1) Correct diffraction of waves. (1)	**2**	Diffraction is the spreading out of waves when passing obstacles. The frequency/wavelength remains unchanged.
6.	(a)	(i)	Count rate measured for fixed time period. (1) Several counts taken at intervals. (1) Plot graph of count rate (vs. time). (1)	**3**	Make sure and mention the apparatus used to take the measurements.
	(a)	(ii)	10 minutes	**2**	Remember, the half-life is the time taken for the activity of radioactive source to fall by half.
	(b)		Rocks/soil/food/internal Any two, 1 mark each.	**2**	

Question			Expected Response	Max mark	Top Tips
	(c)		A helium nucleus	**1**	Alpha particles are the most ionising and therefore have the largest mass.
7.	(a)		Nuclear fission	**1**	Fission is the breaking apart of a heavier nucleus into lighter nuclei. Fusion is the joining together of lighter nuclei into a heavier nucleus.
	(b)		$H = Dw_R$ (1) $H = 0{\cdot}06 \times 10^{-6} \times 20$ (1) $H = 1{\cdot}2 \times 10^{-6}$ Sv (1)	**3**	Remember the prefix μ is micro = $\times 10^{-6}$.
	(c)		Mass = $387{\cdot}497 \times 10^{-27} - 386{\cdot}822 \times 10^{-27}$ (1) $= 6{\cdot}75 \times 10^{-28}$ (kg) (1) $E = mc^2$ $E = 6{\cdot}75 \times 10^{-28} \times (3 \times 10^8)^2$ (1) $E = 6{\cdot}1 \times 10^{-11}$ J (1)	**4**	The National 5 exam will include unfamiliar equations. The question will explain clearly the quantities given in the equation so read the question carefully. $6{\cdot}075 \times 10^{-11}$ J acceptable.
8.	(a)		$a = \frac{v - u}{t}$ (1) $a = \frac{12 - 0}{60}$ (1) $a = 0{\cdot}2$ m s^{-2} (1)	**3**	Remember that 'v' is the final velocity and 'u' is the initial velocity. Think: u comes before v in the alphabet so u comes first.
	(b)		Displacement = area under graph (1) $= \left(\frac{1}{2} \times 60 \times 12\right) + (60 \times 12)$ (1) $= 360 + 720$ $= 1080$ m (1)	**3**	The most common mistake here is to use $s = vt$. This can't be used as there is a changing velocity.
	(c)		$E_K = \frac{1}{2}mv^2$ (1) $E_K = \frac{1}{2} \times 100 \times (12)^2$ (1) $E_K = 7200$ J (1)	**3**	Don't forget to square the velocity.

<table>
<tr><th colspan="3">Question</th><th>Expected Response</th><th>Max mark</th><th>Top Tips</th></tr>
<tr><td>9.</td><td>(a)</td><td></td><td>$a = \frac{F}{m}$ (1)
$a = \frac{4{\cdot}6 \times 10^3}{30\,000}$ (1)
$a = 0{\cdot}15$ m s^{-2} (1)</td><td>3</td><td>The 'F' in $F = ma$ is the unbalanced force. This is the resultant of all forces acting on the object. If there is more than one force acting then you must calculate the unbalanced force before using $F = ma$.</td></tr>
<tr><td></td><td>(b)</td><td>(i)</td><td>air resistance
weight/force of gravity
Arrow pointing upwards with label 'air resistance'; arrow pointing downwards with label 'weight/force of gravity', (1) mark each.</td><td>2</td><td>When drawing force diagrams, always show the direction of the force with an arrow. Don't try and list every force you can think of, the question usually only requires two or three forces.</td></tr>
<tr><td></td><td>(b)</td><td>(ii)</td><td>$v = u + at$ (1)
$= 0 + 9{\cdot}8 \times 24$ (1)
$= 235{\cdot}2$ m s^{-1} (1)</td><td>3</td><td>Remember that all objects travelling vertically through the air are subject to an acceleration of 9·8 m s^{-2} on Earth.</td></tr>
<tr><td></td><td>(b)</td><td>(iii)</td><td>Air resistance/frictional force acting against the motion of the parachutist, (1); reduces the unbalanced force/acceleration in the downwards direction, (1).</td><td>2</td><td>The speed calculated in (b) part (ii) is nearly 530 mph. This would be the speed if there was no friction!</td></tr>
<tr><td></td><td>(c)</td><td></td><td>Terminal velocity is when the resultant force acting on the object is zero.</td><td>1</td><td>Every object has a terminal velocity, which is the maximum speed it can reach. At this point, the forces acting on the object are balanced and it therefore travels at a constant speed.</td></tr>
</table>

Question			Expected Response	Max mark	Top Tips
	(d)		4·8 m s^{-1}; 2·2 m s^{-1}; resultant velocity = 5·3 m s^{-1}; north; Scale: 1 cm = 0.5 m s^{-1} (for example) By scale diagram: **1 mark:** correct diagram to scale, length and angle. **1 mark:** correct drawing of resultant (arrow required). **1 mark:** velocity within tolerance $v = 5{\cdot}3 \pm 0{\cdot}3$ m s^{-1}. **1 mark:** bearing within tolerance 115 ± 3 (or $25 \pm 3°$ S of E). By calculation: $v^2 = 4{\cdot}8^2 + 2{\cdot}2^2$ $= 5{\cdot}28$ m s^{-1} $\tan x = 0{\cdot}46$ bearing = 115 (or 25° S of E)	4	Questions on vector diagrams will usually say 'by scale diagram, or otherwise'. At National 5 level, all vector diagrams are right-angled triangles so there is no need to construct a scale diagram. Use Pythagoras to find the resultant velocity ($a^2 = b^2 + c^2$) and then use trigonometry to find the angle (SOH CAH TOA). Vectors have both magnitude and direction so it is important to make a final statement giving the size of the velocity and the angle of direction. It is not good enough to give the angle alone, a direction, e.g. N of W, must be given for the angle.
10.	(a)		An object that orbits a star.	1	There are several definitions that need to be learned here as well as a planet; a star, a moon, a solar system, an exoplanet, a galaxy.
	(b)		Distance = $1900 \times 10^{10} \times 2$ $= 3{\cdot}8 \times 10^{13}$ (m) (1) $t = \frac{d}{v}$ (1) $t = \frac{3{\cdot}8 \times 10^{13}}{3 \times 10^{8}}$ (1) Total time = 126966·67 s (1)	4	The distance to *Voyager* **and back** must be used. Radio signals travel at 3×10^8 m s^{-1}. To find the total time, 300 s (5 minutes) must be added to the calculated value for time.

Question			Expected Response	Max mark	Top Tips
	(c)		$\frac{1900 \times 10^{10}}{149\,597\,871\,000}$ (1) $= 127$ AU (1)	**2**	Although it is an unfamiliar context, reading the question carefully should enable you to work out the correct answer.
11.			Demonstrates no understanding, 0 marks. Demonstrates limited understanding, 1 mark. Demonstrates reasonable understanding, 2 marks. Demonstrates good understanding, 3 marks. **1 mark:** The student has demonstrated a limited understanding of the physics involved. The student has made some statement (s) which is/are relevant to the situation, showing that at least a little of the physics within the problem is understood. **2 marks:** The student has demonstrated a reasonable understanding of the physics involved. The student makes some statement (s) which is/are relevant to the situation, showing that the problem is understood.	**3**	Open-ended question. Try and write down everything you know about the physics described in the question and if possible, include a physics law or equation.

Question			Expected Response	Max mark	Top Tips
			3 marks: The maximum available mark would be awarded to a student who has demonstrated a good understanding of the physics involved. The student shows a good comprehension of the physics of the situation and has provided a logically correct answer to the question posed. This type of response might include a statement of the principles involved, a relationship or an equation, and the application of these to respond to the problem. This does not mean the answer has to be what might be termed an 'excellent' answer or a 'complete' one.		

Practice Exam B

Section 1

Question	Response	Mark	Top Tips
1.	C	1	Think about what conductors have that insulators do not.
2.	D	1	The current in a conductor is directly proportional to the voltage applied.
3.	C	1	To calculate the voltage across a resistor you must know the current in that resistor.
4.	E	1	The total current in a parallel circuit is equal to the sum of the currents in the branches of the parallel circuit.
5.	E	1	Each of the equations can be used to calculate power. Which one you use depends on the information you have available.
6.	C	1	The outward force comes from the additional pressure created within the aircraft.
7.	A	1	Always make sure that you convert temperatures to Kelvin by adding 273.
8.	A	1	Only an increase in temperature will cause the particles to move faster and collide with the container walls with greater force.
9.	C	1	Most waves are transverse waves. All the waves in the electromagnetic spectrum are transverse. Sound vibrations travel differently as longitudinal waves.
10.	B	1	The length indicated in the diagram is two wavelengths so the distance must be halved to find the length of one wave.
11.	E	1	A signal of a shorter wavelength has a higher frequency.
12.	A	1	Angles are always measured from the normal line to the ray of light.
13.	A	1	**P**rotons = **p**ositive, **neutr**ons = **neutr**al.
14.	C	1	Take some time to learn prefixes. They are always multiples of the number 3.

Question	Response	Mark	Top Tips
15.	D	1	Nuclear fusion is when atoms are 'fused' together. Nuclear fission is when atoms are broken apart. Both involve the release of energy.
16.	D	1	Speed, distance and time are all scalar quantities, which means that they can be expressed with size (or magnitude) only.
17.	D	1	Distance is the total distance covered whereas displacement is how far you are from your starting position.
18.	B	1	The distance travelled is the area under the speed-time graph.
19.	B	1	Mass is the amount of matter in an object and does not change from planet to planet. Weight depends on the gravitational field strength of a planet and can therefore change.
20.	D	1	If an object is travelling at a constant speed then the forces acting on it are balanced.

Practice Exam B

Section 2

Question			Expected response	Max mark	Top Tips
1.	(a)		$E_P = mgh$ (1) $E_P = 70 \times 9{\cdot}8 \times 9$ (1) $E_K = 6174$ J (1)	**4**	Remember that g on Earth is $9{\cdot}8$ m s^{-2}.
	(b)	(i)	Y	**1**	Kinetic energy will be greatest at the lowest point, just as potential energy will be the greatest when the skateboarder and board are at the highest point.
	(b)	(ii)	$E_k = \frac{1}{2} mv^2$ (1) $E_k = \frac{1}{2} \times 70 \times (5{\cdot}5)^2$ (1) $E_k = 1058{\cdot}75$ J (1)	**5**	The law of conservation of energy says that 'energy cannot be created or destroyed, only transferred from one form to another'. In this case it means the kinetic energy the

Question			Expected response	Max mark	Top Tips
			$E_p = mgh$ $1058{\cdot}75 = 70 \times 9{\cdot}8 \times h$ (1) $h = 1{\cdot}54$ m (1)		skateboarder has when moving is equal to the potential energy they have when they reach their maximum height from point Z.
2.	(a)		$E = Pt$ (1) $E = 2000 \times 40$ (1) $E = 80\ 000$ J (1)	**3**	Don't forget the standard units for power are Watts (W). 1 kW = 1000 W.
	(b)		$\Delta T = (84 - 25)$°C $= 59$ °C (1) $E_h = c\,m\,\Delta T$ (1) $E_h = 4180 \times 0{\cdot}3 \times 59$ (1) $E_h = 73\ 986$ J (1)	**4**	ΔT represents the **change in** temperature. The specific heat capacity for water can be found in the data sheet at the front of the exam paper. Make sure you are familiar with the information that is available in the data sheet.
	(c)		Not all electrical energy converted to heat energy. (1) Some energy is lost to surroundings/other parts of the kettle. (1)	**2**	You must give some indication of where energy is lost to.
3.	(a)		$V_{\text{variable resistor}} = 4{\cdot}3$ V (1) $\frac{V_1}{V_1} = \frac{R_1}{R_2}$ (1) $\frac{0{\cdot}7}{4{\cdot}3} = \frac{R_1}{8600}$ (1) $R_1 = 1400\ \Omega$ (1)	**4**	Alternative method using $V = IR$: $V_{\text{variable res}} = IR$ (1) $4{\cdot}3 = 8600$ I $I = 5 \times 10^{-4}$ A (1) $V_{LDR} =$ IR $0{\cdot}7 = 5 \times 10^{-4}$ R (1) $R = 1400\ \Omega$ (1)
	(b)	(i)	Resistance decreases	**1**	Try and use words like 'increases' and 'decreases' instead of 'bigger' or 'smaller'.
	(b)	(ii)	20 lux	**1**	
	(c)	(i)	NPN Transistor	**1**	There are several circuit symbols that need to be learned: a cell, a battery, a lamp, a switch, a resistor, a variable resistor, a voltmeter, an ammeter, an LED, a motor, a microphone, a loudspeaker, a photovoltaic cell, a fuse, a diode, a capacitor, a thermistor, an LDR, a relay.

Question			Expected response	Max mark	Top Tips
	(d)	(ii)	(As light increases) input voltage to transistor increases, (1); (above 0·7 V) switching transistor on, (1); (relay) switch closes/activates, (1).	3	This type of question is built around the transistor switching on. Current will not flow in the street light circuit until the transistor switches on. Your answer must explain the circumstances required to switch the transistor on and also give a description of what happens to the output device once current flows.
4.	(a)		$E = hf_0$ $E = 6{\cdot}63 \times 10^{-34} \times 1{\cdot}18 \times 10^{15}$ (1) $E = 7{\cdot}82 \times 10^{-19}$ (J) (1) Gold (1)	3	The National 5 exam will include unfamiliar equations. The question will explain clearly the quantities given in the equation so read the question carefully.
	(b)		$v = f\lambda$ (1) $3 \times 10^8 = 4{\cdot}8 \times 10^{14}\,\lambda$ (1) $\lambda = 6{\cdot}25 \times 10^{-7}$ m (1)	3	Longest wavelength means highest frequency, in other words the threshold frequency for potassium. This can be calculated by dividing the work function for potassium by Planck's constant. The speed of light in air is 3×10^8 m s^{-1}. This can be found in the data sheet at the front of the exam paper. Make sure you are familiar with the information that is available in the data sheet.
5.			Demonstrates no understanding, 0 marks. Demonstrates limited understanding, 1 mark. Demonstrates reasonable understanding, 2 marks.	3	Open-ended question. Try and write down everything you know about the physics described in the question and if possible, include a physics law or equation.

Question			Expected response	Max mark	Top Tips
5.			Demonstrates good understanding, 3 marks. **1 mark**: The student has demonstrated a limited understanding of the physics involved. The student has made some statement (s) which is/are relevant to the situation, showing that at least a little of the physics within the problem is understood. **2 marks**: The student has demonstrated a reasonable understanding of the physics involved. The student makes some statement (s) which is/are relevant to the situation, showing that the problem is understood. **3 marks**: The maximum available mark would be awarded to a student who has demonstrated a good understanding of the physics involved. The student shows a good comprehension of the physics of the situation and has provided a logically correct answer to the question posed. This type of response might include a statement of the principles involved, a relationship or an equation, and the application of these to respond to the problem. This does not mean the answer has to be what might be termed an 'excellent' answer or a 'complete' one.		Open-ended question. Try and write down everything you know about the physics described in the question and if possible, include a physics law or equation.

Question			Expected response	Max mark	Top Tips
6.	(a)		3×10^{8} m s^{-1}	**1**	Remember that all signals in the electromagnetic spectrum travel at 3×10^{8} m s^{-1}. No units = no marks.
	(b)		$d = vt$ (1) $9{\cdot}46 \times 10^{18} = 3 \times 10^{8} t$ (1) $t = 3{\cdot}15 \times 10^{10}$ s (1)	**3**	Always write down the equation directly from the relationships sheet before re-arranging. This will always guarantee at least 1 mark.
	(c)		$f = \frac{1}{T}$ (1) $f = \frac{1}{6{\cdot}2 \times 10^{-3}}$ (1) $f = 161{\cdot}29$ Hz (1)	**3**	Here, the number of rotations in one second is the frequency. The prefix 'm' for milli $= \times 10^{-3}$.
	(d)		$P = 0{\cdot}033 + 0{\cdot}000013$ T $P = 0{\cdot}033 + 0{\cdot}000013 \times 10\,000$ (1) $P = 0{\cdot}163$ s (1)	**2**	Read the question carefully as all quantities in the unfamiliar equation are defined.
7.	(a)		normal / air / glass / infrared ray	**2**	When light passes from air into glass, it bends towards the normal line.
	(b)	(i)	X-rays	**1**	Learn the sequence of the electromagnetic spectrum. Try making up a rhyme using the first letter of each signal.
	(b)	(ii)	Gamma radiation	**1**	The higher the frequency, the greater the energy the wave has.
	(b)	(iii)	All travel at the same speed/all exhibit interference.	**1**	All travel at 3×10^{8} m s^{-1}.
8.	(a)		Mass that is radioactive $= 0{\cdot}2$ kg (1) $0{\cdot}2$ kg $= 20\%$ of 1 kg (1) $\frac{1{\cdot}3 \times 10^{8}}{5}$ (1) $= 2{\cdot}6 \times 10^{7}$ Bq (1)	**4**	You first need to calculate how much Rutherfordine is contained in the 0·4 kg sample.

Question			Expected response	Max mark	Top Tips
	(b)		Gamma: $H = Dw_R$ $H = 0{\cdot}1 \times 10^{-6} \times 1$ $H = 0{\cdot}1 \times 10^{-6}$ Sv (1) Fast neutrons: $H = Dw_R$ $H = 1 \times 10^{-6} \times 10$ $H = 1 \times 10^{-5}$ Sv (1) Slow neutrons: $H = Dw_R$ $H = 4 \times 10^{-6} \times 3$ $H = 1{\cdot}2 \times 10^{-5}$ Sv (1) $0{\cdot}1 \times 10^{-6} + 1 \times 10^{-5} + 1{\cdot}2 \times 10^{-5}$ $= 2{\cdot}21 \times 10^{-5}$ Sv per hour (1) $2{\cdot}21 \times 10^{-5} \times 8 = 1{\cdot}768 \times 10^{-4}$ Sv (1)	**5**	Equivalent dose calculations must be performed separately for each type of radiation. Calculate the total equivalent dose per hour. The limit is for an 8-hour working day.
9.	(a)		Constant horizontal velocity. (1) Constant vertical acceleration. (1)	**2**	This is the textbook explanation for projectile motion.
	(b)		$d = vt$ (1) $5{\cdot}8 = 12\,t$ (1) $t = 0{\cdot}48$ s (1)	**3**	Horizontal velocity is constant. In other words the ball travels at 12 m s^{-1} horizontally for the whole 5·8 m.
	(c)		$v = u + at$ (1) $v = 0 + 9{\cdot}8 \times 0{\cdot}48$ (1) $v = 4{\cdot}7$ m s^{-1} (1)	**3**	The ball is subject to a constant acceleration due to gravity of 9·8 m s^{-2}. Because it is launched horizontally, the initial vertical velocity is zero. $v = u + at$ is a re-arrangement of $a = v - \frac{u}{t}$? that is useful when performing projectile calculations.
	(d)		speed/m s^{-1} 4.7 0.48 time/s	**2**	Vertically the projectile is subject to a constant acceleration. The graph therefore has a constant gradient.

Question			Expected response	Max mark	Top Tips
10.	(a)	(i)	$W = mg$ (1) $W = 2{\cdot}8 \times 10^6 \times 1{\cdot}6$ (1) $W = 4480000$ N (1)	**3**	The value for the gravitational field strength of the Moon can be found in the data sheet at the front of the exam paper. Make sure you are familiar with the information that is available in the data sheet.
	(a)	(ii)	engine force/thrust (arrow up) weight/force of gravity (arrow down)	**2**	When drawing force diagrams, always show the direction of the force with an arrow. Don't try and list every force you can think of, the question usually only requires two or three forces. 'Gravity' is not a force. You must use either 'weight' or 'force of gravity'.
	(a)	(iii)	$F = 3{\cdot}5 \times 10^7 - 4480000$ $F = 30520000$ N (1) $F = \text{ma}$ (1) $30520000 = 2{\cdot}8 \times 10^6\ \text{a}$ (1) $\text{a} = 10{\cdot}9\ \text{m s}^{-2}$ (1)	**4**	The F in $F = ma$ is the unbalanced force. This is the resultant of all forces acting on the object. If there is more than one force acting then you must calculate the unbalanced force before using $F = ma$.
	(b)		$W = mg$ $W = 2{\cdot}8 \times 10^6 \times 9{\cdot}8$ $W = 27440000$ N (1) $F = 3{\cdot}5 \times 10^7 - 27440000$ $F = 7560000$ N (1) $F = ma$ $7560000 = 2{\cdot}8 \times 10^6\ \text{a}$ $a = 2{\cdot}7\ \text{m s}^{-2}$ (1) Acceleration is less as $2{\cdot}7\ \text{m s}^{-2} < 10{\cdot}9\ \text{m s}^{-2}$ (1)	**4**	The value for the gravitational field strength of the Earth can be found in the data sheet at the front of the exam paper. Make sure you are familiar with the information that is available in the data sheet. Remember that F is the **unbalanced force.**

Question			Expected response	Max mark	Top Tips
11.	(a)		Spectroscope	**1**	Read the passage carefully.
	(b)		The intensity of the absorption lines; the darker the line, the more of the element is present.	**1**	Read the passage carefully.
	(c)		Sodium, hydrogen and helium.	**2**	Match up the lines from the Sun with the lines from the given elements.
12.			Demonstrates no understanding, 0 marks. Demonstrates limited understanding, 1 mark. Demonstrates reasonable understanding, 2 marks. Demonstrates good understanding, 3 marks. **1 mark**: The student has demonstrated a limited understanding of the physics involved. The student has made some statement (s) which is/are relevant to the situation, showing that at least a little of the physics within the problem is understood. **2 marks**: The student has demonstrated a reasonable understanding of the physics involved. The student makes some statement (s) which is/are relevant to the situation, showing that the problem is understood.	**3**	Open-ended question. Try and write down everything you know about the physics described in the question and if possible, include a physics law or equation.

Question			Expected response	Max mark	Top Tips
			3 marks: The maximum available mark would be awarded to a student who has demonstrated a good understanding of the physics involved. The student shows a good comprehension of the physics of the situation and has provided a logically correct answer to the question posed. This type of response might include a statement of the principles involved, a relationship or an equation, and the application of these to respond to the problem. This does not mean the answer has to be what might be termed an 'excellent' answer or a 'complete' one.		

Practice Exam C

Section 1

Question	Response	Mark	Top Tips
1.	C	1	Conservation of energy needs to be applied here. The work done is equal to the kinetic energy produced.
2.	A	1	Charge = current × time. The symbol Q represents charge.
3.	C	1	Think about the quantities involved in the specific heat capacity equation – mass and temperature.
4.	B	1	There are two transistor symbols to learn; the NPN transistor and the *n*-channel enhancement MOSFET.
5.	C	1	There are three power equations that can be used in circuits involving voltage, current and resistance. Select the correct one from the quantities given in the question.
6.	D	1	In a parallel circuit, every time a resistor is added the total resistance reduces.
7.	C	1	You need to calculate the downward force acting on the elephant (weight). This force is distributed across all four feet.
8.	C	1	Check the equation sheet for National 5 Physics. The gas laws that involve a division are directly proportional. A multiplication is inverse proportion.
9.	E	1	Make sure you know your prefixes for every exam. 'm' is the prefix 'milli' which is $\times 10^{-3}$.
10.	C	1	**Long**itudinal waves vibrate a**long** the wave.
11.	E	1	Try and come up with a mnemonic to help you remember the correct order of the electromagnetic spectrum.
12.	D	1	When light moves from a more dense material to a less dense material, the light bends away from the normal line.
13.	A	1	Angles on ray diagrams are always measured from the normal line to the ray of light.
14.	C	1	Alpha and beta radiation are both made of particles but gamma radiation is a wave of energy.

Question	Response	Mark	Top Tips
15.	D	1	When you are asked to give units for a quantity, look at the relevant equation. Absorbed dose is equal to energy divided by mass. What are the units for energy and mass?
16.	D	1	Vector quantities have magnitude (size) and direction whereas scalars have magnitude only.
17.	D	1	An acceleration of 6 m s^{-2} means that the object gains 6 m s^{-1} of speed every second.
18.	A	1	If more than one force is acting on a object then you must calculate the difference between these forces. This is called the resultant or unbalanced force.
19.	B	1	The horizontal speed of a projectile is constant throughout its time of flight.
20.	A	1	Remember Newton's 3rd Law – when A exerts a force on B, B exerts an equal and opposite force on A.

Practice Exam C

Section 2

Question			Expected response	Max mark	Top Tips
1.	(a)	(i)	$\frac{1}{R_t}=\frac{1}{R_2}+\frac{1}{R_3}+\frac{1}{R_3}$ (1) $\frac{1}{R_t}=\frac{1}{10}+\frac{1}{10}+\frac{1}{20}$ (1) $\frac{1}{R_t}=\frac{5}{20}$ $R_t=\frac{20}{5}$ $R_t=4\ \Omega$ (1)	**3**	You cannot simply add the resistor values together. This can only be done with resistors in series. When adding fractions be sure to find the lowest common denominator.
	(a)	(ii)	$I_S=0{\cdot}5$ A $V=I_S R_t$ (1) $V=0{\cdot}5\times 4$ (1) $V=2$ V (1)	**3**	The total current from the battery is 0·5 A. Use this with total resistance calculated in (a) part (i) to find V_S.

Question			Expected response	Max mark	Top Tips
	(b)		$V = IR$ $2 = 20\ I$ (1) $I = 0{\cdot}1$ A (1) $Q = It$ (1) $Q = 0{\cdot}1 \times 300$ (1) $Q = 30$ C (1)	**5**	Remember the voltage is the same across all components in a parallel circuit. To calculate the current in R_3 you need to use the voltage across R_3 and the resistance of R_3. Time must be in seconds.
2.	(a)		$E_k \text{lost} = \frac{1}{2} mv^2$ (1) $E_k \text{lost} = \frac{1}{2} \times 150 \times (1400)^2$ (1) $E_k \text{lost} = 147\,000\,000$ J (1) $E_h = cm\,\Delta T$ (1) $147\,000\,000 = 1050 \times 150 \times \Delta T$ (1) $\Delta T = 933.3$ °C (1)	**6**	Conservation of energy states that energy cannot be created or destroyed, only transformed from one form into another. Here, the kinetic energy is transferred to heat energy. ΔT is the symbol for change in temperature.
	(b)		Less (1) Not all kinetic energy is converted to heat energy. (1)	**2**	Some kinetic energy could be converted to other forms such as sound and light.
3.	(a)		$T_1 = 17 + 273 = 290$ K $T_2 = 69 + 273 = 342$ K (1) $\frac{P_1}{T_1} = \frac{P_2}{T_2}$ (1) $\frac{121}{290} = \frac{P_2}{342}$ (1) $P_2 = 142{\cdot}7$ kPa (1)	**4**	When performing gas laws calculations, temperature must always be converted to Kelvin. None of the gas laws relationships involving temperature will work when using degrees Celsius. 0 °C = 273 K 0 K = –273 °C

Question			Expected response	Max mark	Top Tips
	(b)		More molecules/atoms/particles hit container walls per second. (1) Collisions with walls harder/larger force. (1)	**2**	The kinetic model describes pressure in a container as being caused by the collision of particles with the container walls. An increase in temperature causes those particles to move around faster.
	(c)		Thermometer inside flask of air. (1) Temperature of air measured rather than temperature of water. (1)	**2**	Problem solving question. Check the apparatus diagram. What improvements could be made?
4.	(a)		$n\lambda = d\sin\theta$ $1 \times 700 \times 10^{-9} = 1{\cdot}6 \times 10^{-6} \sin\theta$ (1) $\sin\theta = \frac{700 \times 10^{-9}}{1{\cdot}6 \times 10^{-6}}$ $\sin\theta = 0{\cdot}4375$ (1) $\theta = 25{\cdot}9°$ (1)	**3**	The National 5 exam will include unfamiliar equations. The question will explain clearly the quantities given in the equation so read the question carefully.
	(b)	(i)	$n\lambda = d\sin\theta$ $1 \times 450 \times 10^{-9} = 1.6 \times 10^{-6} \sin\theta$ (1) $\sin\theta = \frac{450 \times 10^{-9}}{1{\cdot}6 \times 10^{-6}}$ $\sin\theta = 0{\cdot}28$ (1) $\theta = 16.3°$ (1)	**3**	Remember the prefix 'n' is nano $= \times 10^{-9}$.
	(b)	(ii)	Angle from zero order maximum is greater for red light than for blue light. (1) Maxima for red light are further apart than those for blue light (1)	**2**	From your answers in (a) and (b) part (i), which has the greater angle? How will this affect the spacing between the maxima?

Question			Expected response	Max mark	Top Tips
5.			Demonstrates no understanding, 0 marks. Demonstrates limited understanding, 1 mark. Demonstrates reasonable understanding, 2 marks. Demonstrates good understanding, 3 marks. **1 mark:** The student has demonstrated a limited understanding of the physics involved. The student has made some statement (s) which is/are relevant to the situation, showing that at least a little of the physics within the problem is understood. **2 marks:** The student has demonstrated a reasonable understanding of the physics involved. The student makes some statement (s) which is/are relevant to the situation, showing that the problem is understood. **3 marks:** The maximum available mark would be awarded to a student who has demonstrated a good understanding of the physics involved. The student shows a good comprehension of the physics of the situation and has provided a logically correct answer to the question posed. This type of response might include a statement of the principles involved, a relationship or an equation, and the application of these to respond to the problem. This does not mean the answer has to be what might be termed an 'excellent' answer or a 'complete' one.	**3**	Open-ended question. Try and write down everything you know about the physics described in the question and, if possible, include a physics law or equation.

Question			Expected response	Max mark	Top Tips
6.	(a)	(i)	0·2 m	**1**	The amplitude is **half** the vertical height of a wave.
	(a)	(ii)	0·4 m	**1**	The wavelength is the distance from one point on a wave to the same point on the next wave.
	(b)	(i)	$f = \frac{45}{60} = 0{\cdot}75$ Hz (1) $v = f\lambda$ (1) $v = 0{\cdot}75 \times 0{\cdot}4$ (1) $v = 0{\cdot}3$ m s^{-1} (1)	**4**	Frequency is the number of waves per second.
	(b)	(ii)	$T = \frac{1}{f}$ (1) $T = \frac{1}{0{\cdot}75}$ (1) $T = 1.33$ s (1)	**3**	The period is the time taken for one wave to pass (in seconds).
7.	(a)	(i)	3×10^8 m s^{-1}	**1**	Remember that all signals in the electromagnetic spectrum travel at 3×10^8 m s^{-1}. No units = no marks.
	(a)	(ii)	$d = vt$ (1) $36000 \times 10^3 = 3 \times 10^8\, t$ (1) $t = 0{\cdot}12$ s (1)	**3**	Don't forget the prefix 'k' is kilo = $\times 10^3$.
	(b)		It moves with constant speed in the horizontal direction. (1) While accelerating due to the force of gravity in the vertical direction. (1)	**2**	A satellite is always falling towards the Earth but never hits the ground due to the curvature of the Earth.

Question			Expected response	Max mark	Top Tips
8.	(a)		$A=\frac{N}{t}$ (1) $850\times10^6=\frac{N}{90}$ (1) $N=7{\cdot}65\times10^{10}$ (1)	**3**	Time must be converted to seconds. No units are required for 'N'.
	(b)		X: 850 → 425 → 212·5 → 106·25 → 53·125 → 26·56 (1) Y: 1800 → 900 → 450 → 225 → 112·5 → 56·25 → 28·125 (1) X ≈ 90 days (1) Y ≈ 36 days (1) Source Y reaches 30 MBq first. (1)	**5**	When performing a series of calculations as in this question, it is good practice to label each individual calculation, e.g. X, Y. This way, the marker can clearly see which radioactive source is first to reach 30 MBq. X will half its activity every 18 days. Y will half its activity every six days.
9.	(a)		Vector quantities have magnitude/size and direction. (1) Scalar quantities have magnitude/size only. (1)	**2**	When describing differences make sure that you clearly describe what is different about **both** items mentioned. An answer of 'vector quantities have direction' would get 0 marks. The answer would have to be 'scalar quantities have magnitude only, vector quantities have magnitude and direction'.

Question			Expected response	Max mark	Top Tips
	(b)	(i)	$a^2 = 700^2 + 800^2$ (1) $a = 1063$ m (1) $\tan\theta = \frac{700}{800}$ (1) $\theta = 41{\cdot}2°$ Direction: 41·2° W of N/48·8° N of W/319. (1) Or by scale drawing: 1 mark for correct diagram to scale, length and angle. 1 mark for correct drawing of resultant (arrow required). 1 mark for displacement within tolerance $v = 1063 \pm 3$ m. 1 mark for bearing within tolerance 319 ± 3 (or $41 \pm 3°$ W of N).	**4**	Questions on vector diagrams will usually say 'by scale diagram, or otherwise..'. At National 5 level, all vector diagrams are right-angled triangles so there is no need to construct a scale diagram. Use Pythagoras to find the resultant velocity ($a^2 = b^2 + c^2$) and then use trigonometry to find the angle (SOH CAH TOA). Vectors have both magnitude and direction so it is important to make a final statement giving the size of the velocity and the angle of direction. It is not good enough to give the angle alone, a direction, e.g. N of W, must be given for the angle.
	(b)	(ii)	$v = \frac{s}{t}$ (1) $v = \frac{1063}{720}$ (1) $v = 1{\cdot}5$ m s^{-1} (1)	**3**	's' is the symbol for displacement. When calculating velocity, displacement must be used, as both quantities are vectors.
	(c)		$t = \frac{d}{v}$ (1) $t = \frac{1500}{1{\cdot}9}$ (1) $t = 789{\cdot}5$ s (1) Sarah arrives first as $720\text{ s} < 789{\cdot}5\text{ s}$. (1)	**4**	Speed and distance are scalars so can be used together.

Question			Expected response	Max mark	Top Tips
10.	(a)		$a = \frac{v-u}{t}$ (1) $a = \frac{17-0}{20}$ (1) $a = 0{\cdot}85\ \text{m s}^{-2}$ (1)	**3**	Remember that 'v' is the final velocity and 'u' is the initial velocity. Think: **u** comes before **v** in the alphabet so **u** comes first.
	(b)		Distance = area under graph (1) $= \left(\frac{1}{2} \times 20 \times 17\right)$ (1) $= 170\ \text{m}$ (1)	**3**	The most common mistake here is to use $d = vt$. This can't be used as there is a changing speed.
	(c)	(i)	Tight clothing/crouched position, etc.	**1**	What does the cyclist do to travel faster?
	(c)	(ii)	Between feet and pedals/between chain and gears, etc.	**1**	Friction can be a useful force too!
11.	(a)		$2 \times 10^{30} \times 3$ (1) $= 6 \times 10^{30}\ \text{kg}$ (1)	**2**	Read passage carefully.
	(c)		$v = \sqrt{\frac{2GM}{R}}$ $v = \sqrt{\frac{2 \times 6{\cdot}7 \times 10^{-11} \times 2{\cdot}04 \times 10^{31}}{30 \times 10^{3}}}$ (1) $v = \sqrt{\frac{2{\cdot}7336 \times 10^{21}}{30 \times 10^{3}}}$ $v = \sqrt{9{\cdot}122 \times 10^{16}}$ (1) $v = 3{\cdot}02 \times 10^{8}\ \text{m s}^{-1}$ (1)	**3**	The National 5 exam will include unfamiliar equations. The question will explain clearly the quantities given in the equation so read the question carefully.

Question			Expected response	Max mark	Top Tips
12.			Demonstrates no understanding, 0 marks. Demonstrates limited understanding, 1 mark. Demonstrates reasonable understanding, 2 marks. Demonstrates good understanding, 3 marks. **1 mark:** The student has demonstrated a limited understanding of the physics involved. The student has made some statement (s) which is/are relevant to the situation, showing that at least a little of the physics within the problem is understood. **2 marks:** The student has demonstrated a reasonable understanding of the physics involved. The student makes some statement (s) which is/are relevant to the situation, showing that the problem is understood. **3 marks:** The maximum available mark would be awarded to a student who has demonstrated a good understanding of the physics involved. The student shows a good comprehension of the physics of the situation and has provided a logically correct answer to the question posed. This type of response might include a statement of the principles involved, a relationship or an equation, and the application of these to respond to the problem. This does not mean the answer has to be what might be termed an 'excellent' answer or a 'complete' one.	3	Open-ended question. Try and write down everything you know about the physics described in the question and, if possible, include a physics law or equation.